Telecommunication Systems I

Telecommunication Systems I

A textbook covering the Level I syllabus
of the Technician Education Council

P H Smale

Principal Lecturer in Telecommunication
Coventry Technical College

Pitman

Pitman Publishing Limited
39 Parker Street, London WC2B 5PB

Associated Companies
Copp Clark Ltd, Toronto
Fearon-Pitman Publishers Inc, Belmont, California
Pitman Publishing New Zealand Ltd, Wellington
Pitman Publishing Pty Ltd, Melbourne

© P H Smale 1978

First published in Great Britain 1978

Text set in 10/12 Linotron Times
printed by photolithography and bound in Great Britain at The
Pitman Press, Bath

ISBN 0 273 01123 5

Preface

In September 1977 the long-established City and Guilds of London Institute Telecommunications Technician Course Parts I, II and III (270/271) began to be phased out by the introduction of the Technician Education Council (TEC) Certificate in Telecommunications, followed by the TEC Higher Certificate in Telecommunications. Broadly speaking, the TEC Certificate is of similar level to the CGLI Part I and the first year of Part II (i.e. years 1, 2 and 3) and the Higher Certificate similar to the second year of Part II and Part III (i.e. years 4 and 5).

This means, of course, that there will be no equivalent of Part I (or Intermediate as it used to be called) which seems a pity, since quite a number of students in the past never progressed beyond second year, but nevertheless obtained a recognised qualification.

No doubt many will regret the passing of the CGLI course which has done great service over the years but one of the main criticisms was the fact that from the second year onwards students selected one of a number of specialisms to study together with the compulsory Mathematics and Telecommunication Principles. This tended to leave students sadly lacking in knowledge of telecommunication systems generally, and with no idea where their particular specialism fitted into a system with all the others.

So it is indeed encouraging to see that the new TEC Certificate course is more "system" orientated, with a Level I unit entitled Telecommunication Systems I designed to give first-year students a broad introduction to several important aspects of telecommunications as a whole. My own view is that this concept could have been continued to greater advantage in later units to maintain students interest and knowledge beyond one particular specialism, but the opportunity has been lost, with the Level III specialist units of the TEC course being just as isolating as the old CGLI course.

One has to ask why this Telecommunication Systems I unit is followed by a Level II compulsory half-unit entitled Transmission Systems II, instead of a more logical and desirable Telecommunication Systems II. This means that the Telecommunication Systems I unit is the only one in a 3-level (or 3-year) course which really considers telecommunications on a broad front, and it therefore becomes very important.

At the same time, however, it is very important to remember that many students will embark on Level I of the course just a few weeks perhaps after leaving school, and there is a real danger of attempting to introduce too much detail and jargon in the various aspects included in Telecommunication Systems I.

The other point to be borne in mind is that, before these are fully developed, there must inevitably be some mention of certain electrical and magnetic principles in the Physical Science unit, being studied (one hopes) concurrently. One must be very careful not to fire a student's enthusiasm on the intricacies of television, radio, radar and computers to such a degree that they do not wish to bother themselves with the fundamentals of Ohm's law, simple circuit theory or the laws of magnetism and electro-magnetism.

It seems to me that the approach to the objectives in Telecommunication Systems I must be as simple and direct as possible, with not too much detail in depth that might well overlap second or third level studies. Of course, not everyone will agree with the approach adopted in this book, or the depth of treatment given, but this is my interpretation as a result of many years of teaching telecommunication technicians, and giving a great deal of careful consideration to this new syllabus.

Only time and experience will show the best way to tackle this. In fact one cannot be sure that the unit objectives are the best possible combination anyway, and amendments to this and other units may be seen to be desirable in the light of experience. But a start has to be made somewhere, and I think that most of us concerned with this new course are anxious to get it under way.

Since the TEC unit syllabus has been written in unfamiliar general and specific objective terms, and individual colleges, or groups of colleges, will have submitted their own assessment plans for approval by TEC, it seemed to be inappropriate to include any suggested phase-testing questions, since college lecturers will be following their own plans and ideas.

I am sure that everyone will benefit by a wide interchange of ideas in the early stages, and I would welcome any such interchange by telephone or letter.

I would like to take this opportunity of expressing my appreciation of the help given by all fourteen of my colleagues in the Telecommunications Section at Coventry Technical College during long and often heated discussions on the requirements and implications of the new course. I have drawn liberally on these discussions.

Also, my sincere thanks are due to Geoff Kirk of Old Swan Technical College, Liverpool, for reading the manuscript and for offering numerous useful comments and corrections.

Finally, I wish to thank my wife for the many hours spent in patiently typing the original manuscript from my almost unreadable scribbles.

P.H.S.
Coventry, June 1977

Contents

1 Introduction to Telecommunication

Introduction

The development of civilization as we know it today is largely due to man's ability to exchange information and ideas by the natural senses of sight and hearing, and by the written word using some form of accepted language or code. From the very beginning, man has constantly searched for means of passing information beyond the normal range of human vision and hearing. Everyone is familiar with such methods as Indian smoke signals, beacon fires, and semaphore flag signalling.

It is worth pointing out here that "tele" is derived from the ancient Greek for "at a distance," "phon" means sound or speech, "graph" means writing or drawing. So the following well-known terms have emerged:

Telecommunication – communicating at a distance.
Telephone   – speaking at a distance.
Television   – seeing at a distance.
Telegraph   – writing at a distance.

Telecommunication is, then, the process of passing INFORMATION energy over long distances by electrical means. The information energy is passed to the destination either over suitable insulated conducting wires called TRANSMISSION LINES, or through the atmosphere without the use of wires by a RADIO link. These two methods will be explained later on.

The straightforward use of electrical energy or electricity for everyday tasks is well known; for example electric cookers, electric lights, electric motors, etc. In each of these, electrical energy is converted into another form of energy in order to provide the power to carry out a particular task.

In telecommunication, some form of "information" or "intelligence" energy is changed into electrical energy so that it can be passed to a distant point. At the destination the electrical energy is converted back into its original form. This particular use of electrical energy to convey *information* comes under the general heading of *electronics*. Familiar forms of original information energy are human voice sounds, music, visible moving scenes, still (or non-moving) pictures, and so on.

Basic Requirements of Telecommunication Systems

First of all the original information energy must be converted into electrical form to produce an electronic information SIGNAL. This is achieved by a suitable TRANSDUCER, which is a general term given to any device that converts energy from one form to another when required.

Suppose the electronic signal is now passed to the destination by a *line link*, with the energy travelling at a speed approaching that of light, and at the destination a second transducer converts the electronic signal back into the original energy form, as shown in Fig. 1.1. In practical systems other items will be required. For example, AMPLIFIERS may be needed at appropriate points in the system to increase the strength of the electronic signal to acceptable values.

For a *radio* system, a TRANSMITTER is required at the source to send the signal over the radio link without wires, with the energy travelling at the speed of light, and a RECEIVER is required at the destination to recover the signal before applying it to the transducer, as shown in Fig. 1.2.

At this point it is important to realise that, in both these systems, interference will be generated by electronic NOISE, and also that DISTORTION of the electronic signal will occur for a number of reasons. These are undesirable effects and must be minimized in the system design.

It will be obvious from Fig. 1.1 and Fig. 1.2 that these simple systems are one-way or UNIDIRECTIONAL only (generally called a channel), and domestic radio and television broadcasting are familiar examples of such systems.

Other systems, however, such as the national telephone system, must be capable of conveying information in *both* directions. To do this, the basic requirements shown in Fig. 1.1 and Fig. 1.2 must be duplicated in the opposite direction to provide a two-way or BOTHWAY system (generally called a circuit).

Analogue and Coded Signals

Some telecommunication transducers produce an electronic signal that directly follows the instantaneous variations of the original information energy. Such signals are called

SOURCE   DESTINATION

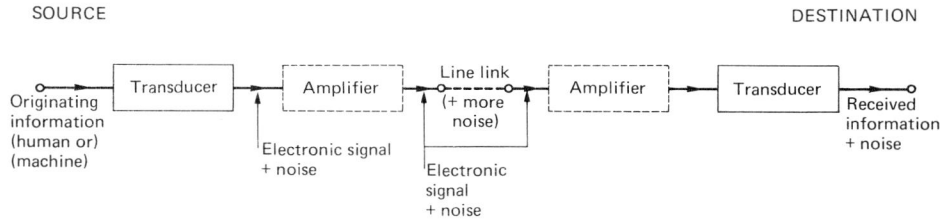

Fig. 1.1 Basic requirements of a one-way line telecommunication channel

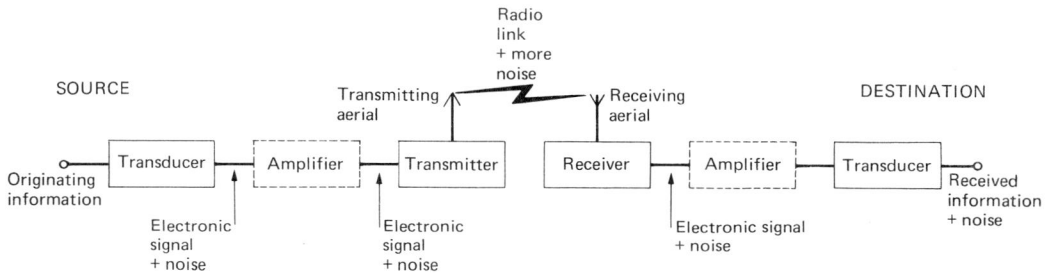

Fig. 1.2 Basic requirements of a one-way radio telecommunication channel

ANALOGUE signals. For example, a *microphone* produces an electronic signal that follows the variations of sound energy that actuate the microphone. A loudspeaker receives the analogue electronic signal and reproduces the original sound energy variations.

There are other systems in which the transducer produces an electronic signal in the form of a pre-determined CODE of pulses or variations that is understood by humans or machines at both ends of the system. One example of this is the teleprinter, which produces a coded electronic signal dependent on which key is depressed on the sending keyboard. The coded signal is passed to the destination where it is accepted by the receiving teleprinter and the appropriate letter or figure is then printed.

Direct and Alternating Currents

In certain electrical circuits the current flows only in *one* direction when the energy supply is connected, although the amount or strength of the current can be controlled. This type of current is known as DIRECT CURRENT (d.c.), and is produced by an energy source such as a dry battery, accumulator or rotating generator.

In other electrical circuits, however, the current reverses direction at regular intervals with a particular repeating pattern or WAVEFORM. This type of current is known as ALTERNATING CURRENT (a.c.), and is generated by an energy source such as a rotating alternator, an electronic oscillator, or certain types of telecommunication transducer.

Use of Direct Current Signals

A steady direct current flowing in a circuit cannot convey information by itself, but the inclusion of a simple on-off switch enables the current to be regulated in a series of pulses. When the switch is opened the current drops to zero, and when the switch is closed the current rises to a steady value. If the current pulses are produced in accordance with a pre-arranged code, whereby each letter or number is represented by a particular combination of pulses, then operation of the switch can send any desired message. The current pulses must actuate a device that enables a second person to "see" or "hear" this message.

The Morse Code is one well-known example of this form of d.c. signalling, and a very simple circuit is shown in Fig. 1.3. Each letter has a code comprising a fixed number of short and long pulses of current called "dots" and "dashes". The complete alphabet code will not be given here, since it is readily available elsewhere but, for illustration, the combination of current pulses representing the letter A is shown in Fig. 1.4.

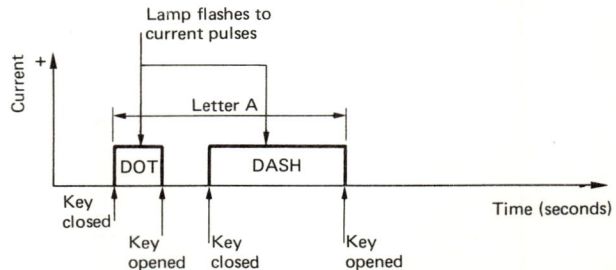

Fig. 1.3 Simple lamp-signalling morse code circuit

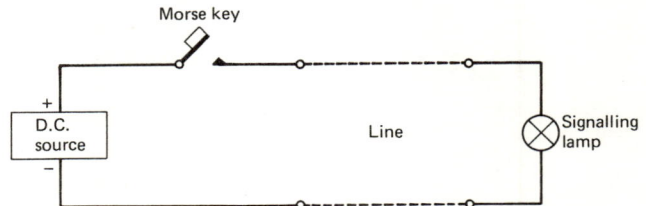

Fig. 1.4 D.C. code representing the letter A in morse code

If in Fig. 1.3 the connections to the d.c. source are reversed, then the direction of current is also reversed. The current direction can therefore be considered as positive or negative, according to which way it is flowing around the circuit, and this is known as the POLARITY of the current.

With this method of d.c. signalling, the information is carried by the alternate absence and presence of the current. It is also possible to convey information by switching the d.c. between two different values. In either case, it is the variation of *amplitude* of the current that is important.

Other typical uses of d.c. signals are

(1) Operating automatic exchange switches from telephone dials.
(2) Control and metering of telephone calls between exchanges.
(3) Operating simple indicating instruments, such as car fuel gauges, etc.

It should be added here that other systems such as p.c.m., data, and ceefax use d.c. signals in various forms, either in the on/off condition, or using current reversals, but in general these will be considered later as a.c. signals.

The main disadvantages of d.c. signals are

(1) Difficulty in transmission over long line circuits due to attenuation and distortion, although regeneration (boosting) and amplification are possible.
(2) Connecting wires are always needed for the whole of a telecommunication circuit.

It is important to realise, however, that sources providing direct current are widely used to supply power or energy to electronic circuits.

Varying or fluctuating d.c. signals have similar characteristics to a.c. signals and will be considered as such later on in the course.

Alternating Current Waveforms

As already explained, alternating currents reverse direction at regular intervals with some repeating pattern or waveform. The main advantages of a.c. signals are

(1) The strength or amplitude can easily be altered (e.g. by transformer, amplifier, etc.), allowing transmission over long lines.
(2) Connecting wires are *not* necessarily required for the *whole* of a telecommunication circuit.

Many waveforms are possible with alternating currents. One of the simplest to produce comes by rotation of a loop of wire in a uniform magnetic field. This is called a SINUSOIDAL waveform and is shown in Fig. 1.5.

Between points A and B the current increases from zero to a peak value in the positive direction. Between points B and C the current gradually reduces to zero. Then, between points C and D, the current "increases" to a peak value in the opposite or negative direction, and between points D and E it gradually "reduces" to zero again. This whole sequence from point A to point E represents one complete rotation of the wire loop in the magnetic field, and is called one CYCLE of a.c. waveform. Clearly the cycle is repeated between points E and F, representing another rotation of the wire loop, and this waveform is repeated for each subsequent rotation.

The time needed, in seconds, for one cycle of waveform to be produced is called the PERIODIC TIME (T) of the a.c. waveform.

The number of complete cycles occuring in one second is called the FREQUENCY (f) of the a.c. waveform in HERTZ (Hz), One hertz is one cycle per second. From Fig. 1.6 it should be clear that frequency and periodic time are reciprocals of each other. That is

$$\text{Frequency} = \frac{1}{\text{Periodic time}} \quad \text{and}$$

$$\text{Periodic time} = \frac{1}{\text{Frequency}}$$

with frequency in hertz (Hz) and time in seconds. From Fig. 1.6, the frequency is 4 Hz and the periodic time is $\frac{1}{4}$ second.

The strength of the current at any instant in time is called the AMPLITUDE of the waveform, and the direction of the current (positive or negative) is called the POLARITY of the current, as in d.c.

It will also be clear from Fig. 1.5 and Fig. 1.6 that the amplitude reaches a PEAK VALUE in the positive *and* negative directions once every cycle.

We have, as one example, already associated the production of a sinusoidal waveform with the rotation of a loop of wire in a magnetic field, and the resulting current plotted against time, as in Fig. 1.5 and Fig. 1.6. We could also consider the loop as moving through 360° in one rotation, so we could plot the resultant current against angular rotation, as shown in Fig. 1.7.

Also, if we consider the energy of an a.c. waveform travelling through space or along a transmission line at a particular velocity, then a certain distance will be travelled in the periodic time for one cycle, as shown in Fig. 1.8. We can see now that the a.c. waveform repeats complete cycles over equal

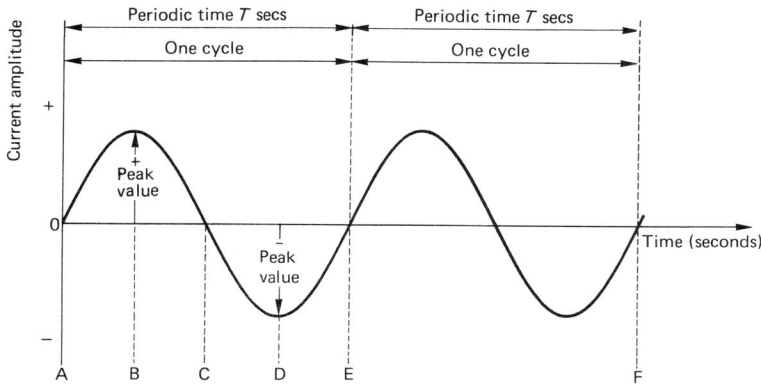

Fig. 1.5 Sinusoidal a.c. waveform

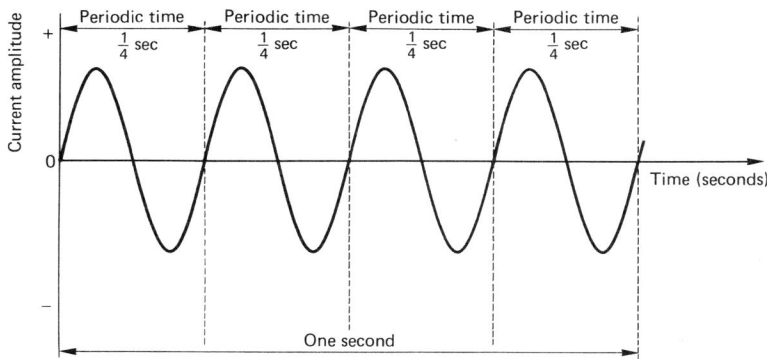

Fig. 1.6 Sinusoidal a.c. waveform with a frequency of 4 Hz

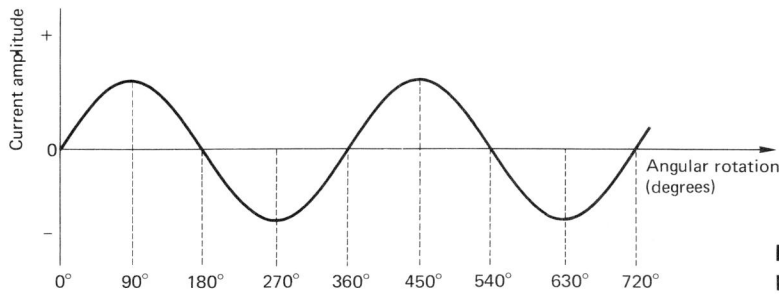

Fig. 1.7 Sinusoidal a.c. waveform plotted against angular rotation

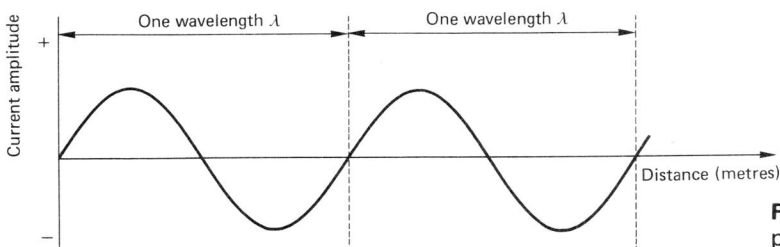

Fig. 1.8 Sinusoidal a.c. waveform plotted against distance

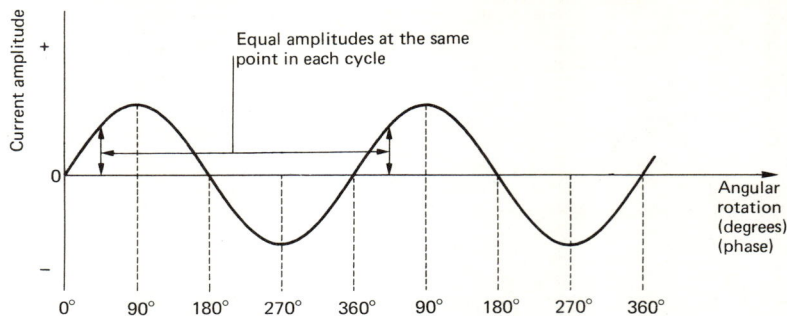

Fig. 1.9 Sinusoidal a.c. waveform related to phase

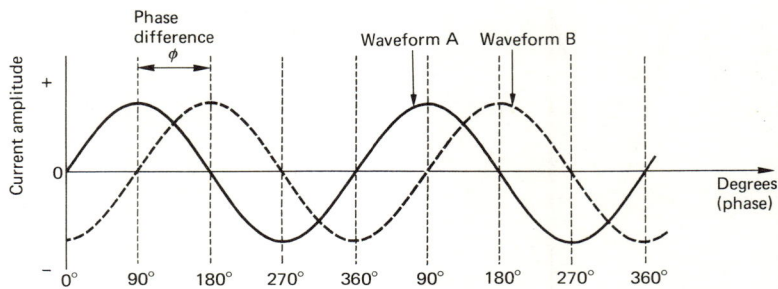

Fig. 1.10 Illustration of phase difference ϕ between two waveforms

distances. The distance representing each cycle is called the WAVELENGTH of the a.c. waveform in metres. The Greek letter lambda (λ) is used as the symbol for wavelength.

In Fig. 1.7 the rotation of the loop is shown as a continuously increasing number of degrees. Alternatively, we could consider the start of each rotation as beginning from 0°, so each successive cycle in fact occurs from 0 to 360°, as shown in Fig. 1.9. Clearly at the *same point* in each cycle the amplitude of the waveform has the *same value*. This way of identifying a particular point in any cycle as a degree of rotation is called the PHASE of the a.c. waveform.

In the same way, if two waveforms are identical *except* for their phase, then the difference between the two can be expressed as a PHASE DIFFERENCE, as shown in Fig. 1.10. In Fig. 1.10, waveform A is seen to be *leading* waveform B by 90°. Put another way, waveform B is *lagging* waveform A by 90°.

Relationship between Frequency, Wavelength and Velocity.

We have seen that an a.c. waveform has a certain energy velocity (metres per second), with a periodic time (T seconds) for the duration of each cycle, and with a certain wavelength distance (λ metres) for each cycle. Now, in general, velocity, distance and time are related by

$$\text{Velocity} = \frac{\text{Distance}}{\text{Time}} \qquad \text{(e.g. metres per second, km/h)}$$

So, for any a.c. waveform of wavelength λ and periodic time T,

$$\text{Velocity } v = \frac{\text{Wavelength } \lambda}{\text{Time } T}$$

But it has already been established that

$$\text{Frequency } f(\text{Hz}) = 1/\text{Periodic time } T(\text{secs})$$

so that

$$\text{Velocity } v = \text{Wavelength } \lambda \times \text{Frequency } f$$

Therefore, for any a.c. waveform, if two of these properties are known, the third can be calculated:

$$v = \lambda f \qquad \lambda = \frac{v}{f} \qquad f = \frac{v}{\lambda}$$

As stated earlier, a.c. energy can be propagated through the atmosphere as a radio wave without the use of wires. This is in fact a type of ELECTROMAGNETIC WAVE that is very similar to light energy, and has the same velocity as light, which is 300 000 000 metres per second and usually indicated by c. Radio waves can be generated over a wide range of frequencies, commencing at around 10 000 Hz and continuing through millions of hertz to thousands of millions of hertz, as indicated in Fig. 1.11. Included also are electromagnetic waves such as visible light, X-rays, etc.

The following abbreviations should be noted:

thousands of hertz, or kilohertz – kHz
millions of hertz, or megahertz – MHz
thousands of millions of hertz, or gigahertz – GHz

The propagation characteristics of radio waves through the atmosphere vary greatly with frequency, and choice of frequency for a particular radio service must take this into account. Radio waves are divided into different frequency bands according to their propagation characteristics, as in Table 1.1.

ELECTROMAGNETIC WAVE SPECTRUM

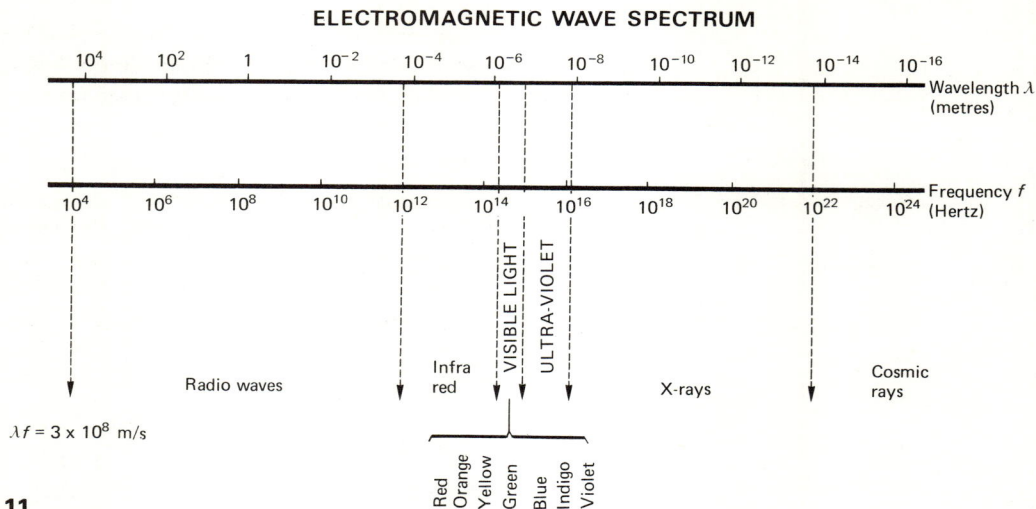

Fig. 1.11

Table 1.1

Frequency band		Corresponding wavelength
Very low frequency (v.l.f.)	below 30 kHz	above 10 000 metres
Low frequency (l.f.)	30 kHz to 300 kHz	10 000 to 1000 m
Medium frequency (m.f.)	300 kHz to 3 MHz	1000 to 100 m
High frequency (h.f.)	3 MHz to 30 MHz	100 to 10 m
Very high frequency (v.h.f.)	30 MHz to 300 MHz	10 to 1 m
Ultra high frequency (u.h.f.)	300 MHz to 3 GHz	1 m to 10 cm
Super high frequency (s.h.f.)	3 GHz to 30 GHz	10 cm to 1 cm
Extremely high frequency (e.h.f.)	above 30 GHz	below 1 cm

Some typical services allocated to the different frequency bands are

V.L.F. Long-distance telegraphy broadcasting.

L.F. Long-distance point-to-point service, navigational aids, sound broadcasting, line carrier systems.

M.F. Sound broadcasting, ship-shore services, line carrier systems.

H.F. Medium and long-distance point-to-point services, sound broadcasting, line carrier systems.

V.H.F.⎫ Short-distance communication, TV and sound broadcasting, radar.

U.H.F.⎭ Air-air and air-ground services.

S.H.F. Point-to-point microwave communication systems, radar.

Knowing the *velocity* of radio waves, then when a certain radio transmission is operating at an allocated *frequency*, the corresponding *wavelength* can be calculated.

EXAMPLE 1.1

The BBC Radio 2 sound programme is received by domestic radio receivers on 1500 metres in the Long Waveband. What is the allocated frequency of the programme?

Now, Velocity $c = 300$ million metres per second, and Wavelength $\lambda = 1500$ metres.
Therefore

$$\text{Frequency } f = \frac{\text{Velocity } c}{\text{Wavelength } \lambda} = \frac{300\,000\,000}{1500}\text{ Hz}$$
$$= 200\,000\,\text{Hz} \quad \text{or} \quad 200\,\text{kHz} \quad (Ans.)$$

EXAMPLE 1.2

If the frequency allocated to Radio Luxembourg is 1.442307 MHz, calculate the corresponding wavelength.

$$\text{Wavelength } \lambda = \frac{\text{Velocity } c}{\text{Frequency (Hz)}}$$
$$= \frac{300\,000\,000}{1\,442\,307}\text{ metres}$$
$$= 208 \text{ metres} \quad (Ans.)$$

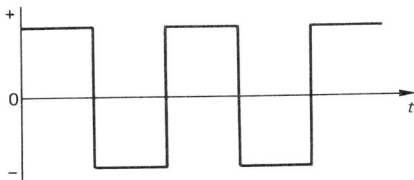

Information-carrying Capacity of A.C. Waveforms

We have already seen that variation in the amplitude of a direct current (for example by switching the d.c. on and off) enables an information signal to be carried by the current. Similarly, an alternating current (or voltage) can be used to carry an information signal from source to destination by arranging for the information signal to vary one of the characteristics of the a.c. waveform – either its amplitude or its frequency or its phase. This will be explained in a later chapter.

Composition of Complex Waveforms

So far, only simple sinusoidal waveforms have been considered, although it was mentioned earlier that many other complex waveforms are possible. Examples of some complex waveforms commonly found in telecommunication are shown in Fig. 1.12.

(a) SQUARE WAVE

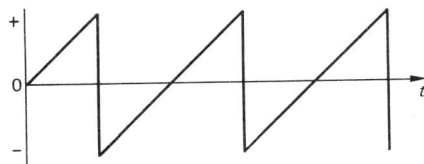

(b) SAW-TOOTH WAVEFORM

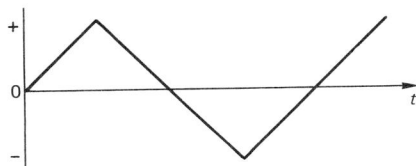

(c) TRIANGULAR WAVEFORM

Fig. 1.12 Typical complex a.c. waveforms

It can be shown by mathematical analysis that any complex waveform is made up of a sinusoidal waveform having a certain frequency called the *fundamental* frequency and a number of other sinusoidal waveforms having frequencies that are direct multiples of the fundamental frequency with decreasing peak values. These direct multiples are called HARMONICS of the fundamental frequency. Some other complex waveforms will contain a d.c. component also, although this is zero for the three waveforms shown in Fig. 1.10.

For example, for a complex waveform having a fundamental frequency f Hz the following harmonics may also be present:

$2f, 3f, 4f, 5f, 6f, \ldots$ etc.

The square wave shown in Fig. 1.12a is made up theoretically of a fundamental frequency f and all the *odd* harmonics rising to infinity, i.e.

$3f, 5f, 7f, \ldots$ etc.

Fig. 1.13 shows how a sinusoidal fundamental waveform and its 3rd and 5th harmonics combine to produce a complex wave, which suggests that a square wave would be produced if further odd harmonics were added.

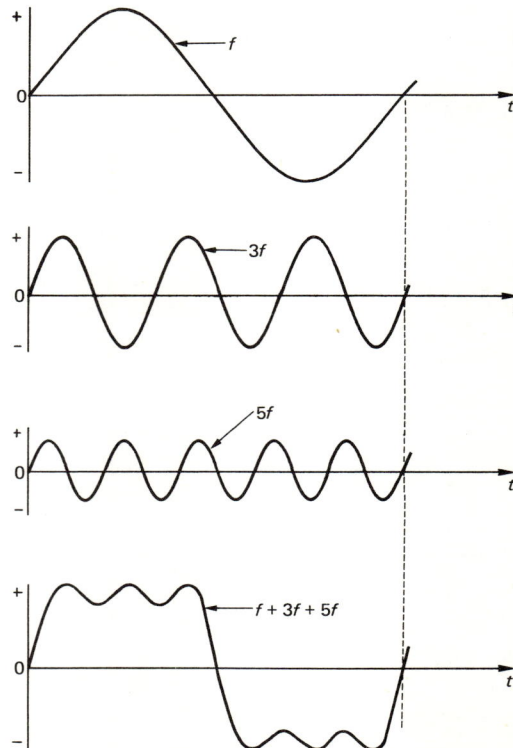

Fig. 1.13 Amplitude addition of fundamental a.c. and its 3rd and 5th harmonics

The saw-tooth waveform shown in Fig. 1.12*b* contains a fundamental sinusoidal waveform and, theoretically, all odd and even harmonics to infinity.

Information Signal Bandwidth

The sound waves produced by a human voice are variations of air pressure above and below normal pressure, so they can be considered as alternating in nature, with a complex waveform that is different for each individual voice. It is this unique nature of the complex waveform of sound waves that enables us to recognise individual voices. Since each voice has a different complex waveform, it must contain certain fundamental frequencies and harmonics. Generally, the range of fundamental frequencies represents the information or intelligence, whilst the harmonic content gives individual recognition. The sound waves produced by human voices therefore must contain a RANGE of frequencies, and the range is known as the BANDWIDTH. Since the microphone produces an electronic signal that is virtually the direct analogue or copy of the sound waves, the speech information electronic signal must also have a minimum frequency bandwidth that must be maintained throughout the system carrying the information.

Other different types of information signal in telecommunication (telegraphy, television, music, data, etc.) have different minimum bandwidths, which will be considered later.

It should be realized that this idea of information signal bandwidth has an important bearing on the design of a system, in terms of circuitry and transmission media.

2 The Modulation Process

Introduction

Chapter 1 introduced the concept of frequency bandwidth of information signals, consisting of fundamental frequencies and related harmonic frequencies. Generally, the fundamental frequencies contain the information and most of the power, and the harmonics give recognition to the original signal source. For example, the number and strength of harmonics present enables individual human voices and different musical instruments to be recognised.

The range of frequencies produced by the average human voice is of the order of 100–7500 Hz. As a simple rule, it can be said that frequency bandwidth transmission costs money, so at a very early stage in the development of national and international telephone networks it was agreed that, in the interests of economy, the natural frequency bandwidth of speech signals produced by talking into a telephone should be restricted to the range 300–3400 Hz, called the COMMERCIAL SPEECH BANDWIDTH. This enables the information to be readily understood by the listener at the distant end without necessarily giving recognition of the talker's voice. In fact, however, it is generally found that the "telephone" voice of any individual becomes recognisable after relatively few conversations.

The range of frequencies produced by a full musical orchestra can be of the order of 30–20 000 Hz, which also is the approximate hearing range of most people with good hearing facilities. For aged people and those with certain hearing defects, the range of frequencies that can be detected will probably be less than this.

In order to enjoy music fully it is essential that the individuality of instruments is not destroyed, so it is not advisable to restrict the natural frequency bandwidth produced. The frequency bandwidth used for music transmission depends on the particular situation, but some common examples are

(*a*) For medium wave radio broadcasting and associated land-links 50–8000 Hz.

(*b*) For BBC V.H.F./F.M. sound broadcasting, 50–15 000 Hz.

(*c*) For high-fidelity sound reproduction (Hi-fi), 30–20 000 Hz.

Line Transmission Systems

At certain points in the national telephone network, the number of telephone calls being handled is very large, and if a separate line were used for every call having a frequency bandwidth of 300–3400 Hz, then very many such lines would be necessary. It is possible to manufacture cables having lines which can handle information signals over very large frequency bandwidths, so it was a natural development to look for ways of allowing many telephone calls to *share* the large bandwidth capacity of these lines.

This technique is called MULTIPLEXING, and the sharing of a line by many telephone channels can be done either on a frequency basis called *frequency-division multiplex*, or on a time basis called *time-division multiplex.*

In order to pass a number of telephone channels simultaneously over a single line, it is necessary to change the original commercial speech bandwidth of 300–3400 Hz of each signal into a completely different band within the frequency capacity of the line. This is achieved by using a different high-frequency waveform to "carry" each individual speech signal. These high-frequency waveforms are called CARRIER frequencies, and the speech information is superimposed on to this carrier to be transmitted to the distant end of the line.

Radio Systems

It was seen in Chapter 1 that it is possible to radiate energy as an electromagnetic wave without the use of wires at frequencies from about 10 kHz upwards. In fact, at very low frequencies (under 30 kHz) it is very expensive to transmit radio waves because of the high power needed from the transmitter, and because the transmitter aerial installation has to be very large. Briefly, in order to radiate energy efficiently, the length of the transmitting aerial needs to approach at least a quarter of a wavelength at the working frequency.

At a frequency of 10 kHz, the aerial would need to be something approaching 7500 metres (around 4.5 miles), since

$$\lambda = c/f = 3 \times 10^8/10^4 \text{ m} = 3 \times 10^4 \text{ m} \qquad \text{i.e. } \lambda/4 = 7500 \text{ m}$$

Clearly this is not very practical.

So, it is very difficult indeed to transmit low-frequency speech and music information signals directly as radio waves. However, at higher frequencies with shorter wavelengths it becomes easier and more economical to transmit radio waves, so radio systems use high frequencies to "carry" the low-frequency information signals to the destination.

In both types of system the information signals are superimposed on to the carrier signal at the transmitting end by a process called MODULATION. At the destination, the information signal is recovered from the carrier by the reverse process called DEMODULATION.

Modulation of a carrier wave is achieved by arranging for some characteristic of the carrier wave to be varied by the information signal. In Chapter 1 it was seen that a sinusoidal a.c. waveform has a number of important characteristics, e.g. peak value (amplitude), frequency, and phase, and it can be arranged for the information signal to vary *any* of these characteristics of the carrier waveform.

In order to simplify the understanding of this modulation process the information signal, or MODULATING SIGNAL, will be considered as a single-frequency waveform instead of a band of frequencies as previously considered. The *modulating signal* and the *carrier wave* are applied to a *modulator* circuit, and the MODULATED CARRIER WAVE is extracted from the output of the modulator circuit, as illustrated in Fig. 2.1. The function of the demodulator will be considered later.

Amplitude Modulation

This is the process of varying the *amplitude* of the sinusoidal carrier wave by the amplitude of the modulating signal, and is illustrated in Fig. 2.2.

The unmodulated carrier wave has a constant peak value and a higher frequency than the modulating signal but, when the modulating signal is applied, the peak value of the carrier varies in accordance with the instantaneous value of the modulating signal, and the outline waveshape or "envelope" of the modulated wave's peak values is the same as the original modulating signal waveshape. The modulating signal waveform has been superimposed on the carrier wave.

Frequency Modulation

This is the process of varying the *frequency* of the sinusoidal carrier wave by the amplitude of the modulating signal, and is illustrated in Fig. 2.3.

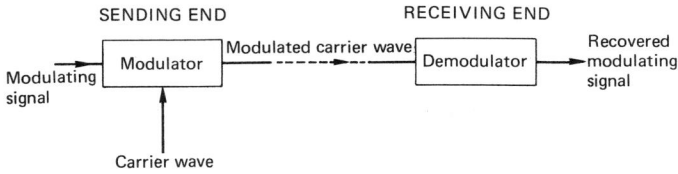

Fig. 2.1 Simple principle of a modulated telecommunication system

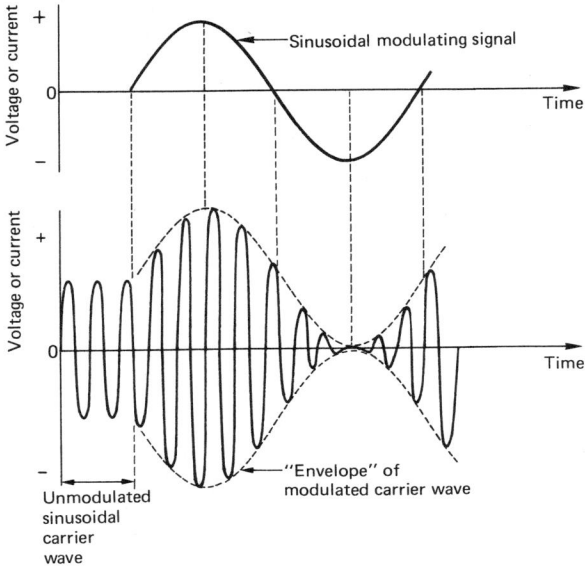

Fig. 2.2 Graphical illustration of an amplitude-modulated carrier wave

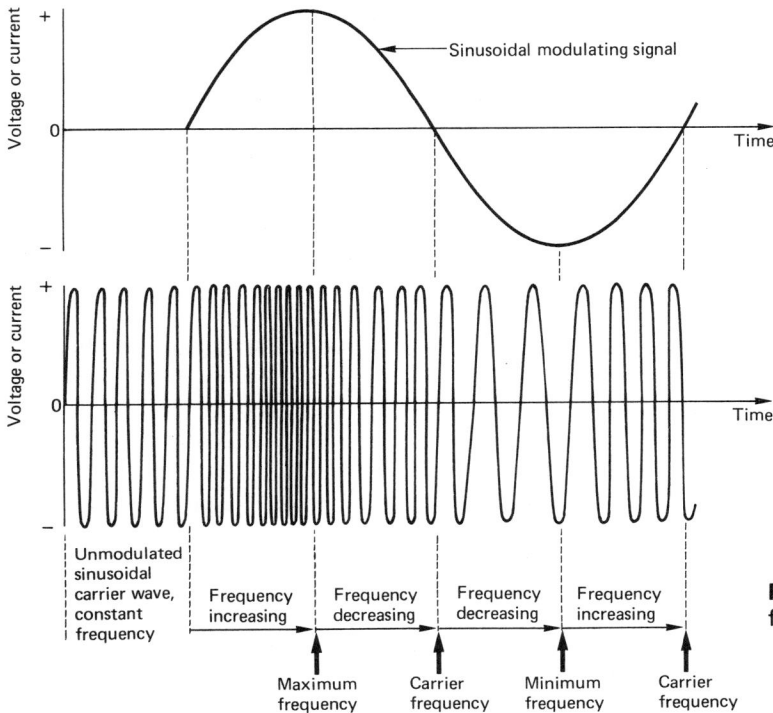

Fig. 2.3 Graphical illustration of a frequency-modulated carrier wave

When the modulating signal is applied, the carrier frequency increases to a maximum value as the modulating signal amplitude increases to a maximum in a positive direction, and decreases to its unmodulated value as the amplitude decreases again towards zero. Then, in the second half-cycle of the modulating signal, the carrier frequency decreases to a minimum value as the modulating signal amplitude increases to a maximum in a negative direction, and increases to its unmodulated value as the modulating signal amplitude decreases again towards zero.

Note that the peak value or amplitude of the carrier wave remains constant. It is important to understand that the variation of the carrier frequency above and below its unmodulated value depends on the *amplitude* of the modulating signal voltage (or current).

Pulse Modulation

Another method of conveying information is by means of pulses of voltage or current.

With pulse modulation the carrier wave used is not sinusoidal, but consists of repeated rectangular pulses. The amplitude, width or position of the pulses can be altered by the amplitude of the information signal, as illustrated in Fig. 2.4.

Bandwidth of Amplitude-modulated Carrier Waves

It can be shown by mathematical analysis that, when a sinusoidal carrier wave of frequency f_c Hz is amplitude-modulated by a sinusoidal modulating signal of frequency f_m Hz, then the modulated carrier wave contains *three* frequencies.

One is the original carrier frequency, f_c Hz.

The second is the *sum* of carrier and modulating signal frequencies,

$$(f_c + f_m) \text{ Hz}$$

The third is the *difference* between carrier and modulating signal frequencies,

$$(f_c - f_m) \text{ Hz}$$

This is illustrated in Fig. 2.5.

It should be noted that two of these frequencies are new, being produced by the amplitude-modulation process, and are called SIDEFREQUENCIES.

The *sum* of carrier and modulating signal frequencies is called the *upper sidefrequency*.

The *difference* between carrier and modulator signal frequencies is called the *lower sidefrequency*.

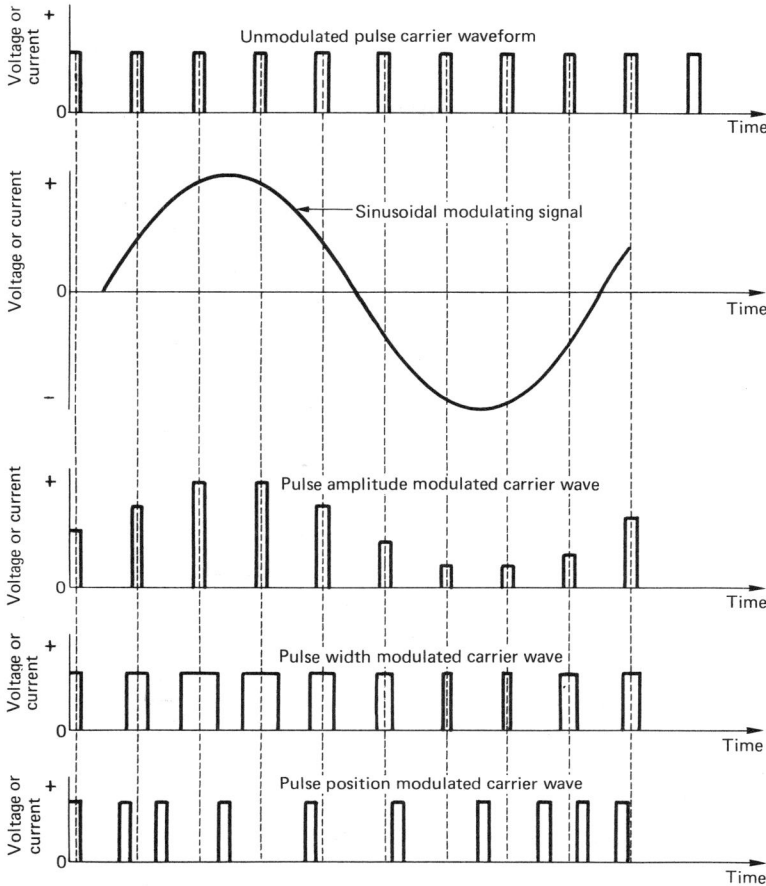

Fig. 2.4 Graphical illustration of a pulse-modulated carrier wave

Unmodulated pulse carrier waveform

Sinusoidal modulating signal

Pulse amplitude modulated carrier wave

Pulse width modulated carrier wave

Pulse position modulated carrier wave

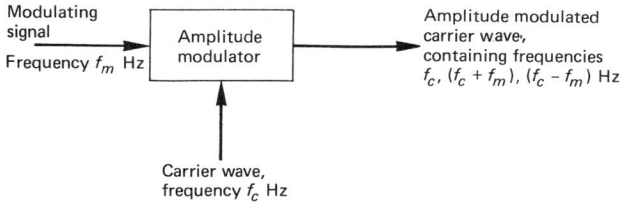

Modulating signal
Frequency f_m Hz

Amplitude modulator

Amplitude modulated carrier wave, containing frequencies f_c, $(f_c + f_m)$, $(f_c - f_m)$ Hz

Carrier wave, frequency f_c Hz

Fig. 2.5 Simple principle of amplitude modulation

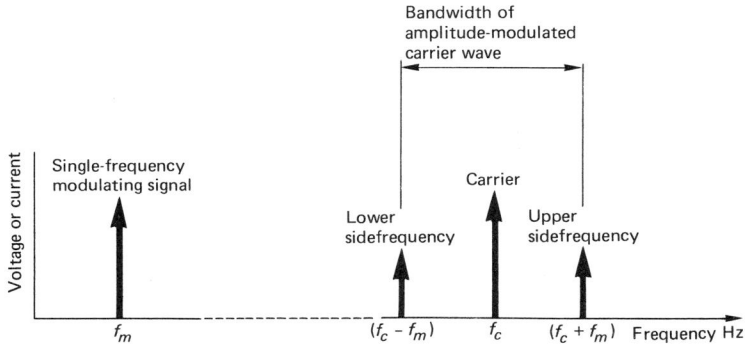

Bandwidth of amplitude-modulated carrier wave

Single-frequency modulating signal

Carrier

Lower sidefrequency

Upper sidefrequency

f_m $(f_c - f_m)$ f_c $(f_c + f_m)$ Frequency Hz

Fig. 2.6 Frequency spectrum of an amplitude-modulated wave for a single-frequency modulating signal

This is illustrated in the frequency spectrum diagram of Fig. 2.6.

The *bandwidth* of the modulated carrier wave is

$$(f_c + f_m) - (f_c - f_m) = 2f_m$$

i.e. *double* the modulating signal frequency.

When the modulating signal consists of a *band* of frequencies, as already seen for commercial speech and music for example, then each individual frequency will produce upper and lower sidefrequencies about the unmodulated carrier frequency, and so upper and lower SIDEBANDS are obtained. This is illustrated in Fig. 2.7.

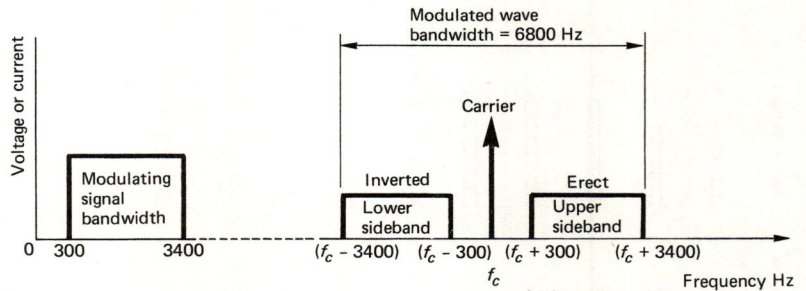

Fig. 2.7 Frequency spectrum of an amplitude-modulated wave for commercial speech modulating signal

The bandwidth of the modulated carrier wave is

$$(f_c + 3400) - (f_c - 3400) = 6400 \text{ Hz}$$

which is *double* the *highest* modulating signal frequency.

It follows therefore that, as the modulating signal bandwidth increases, the modulated wave bandwidth also increases, and the transmission system used must be capable of handling this bandwidth throughout.

3 Introduction to Radio Systems

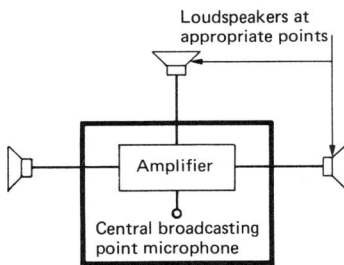

Loudspeakers at appropriate points

Amplifier

Central broadcasting point microphone

Fig. 3.1

Introduction

One important system used extensively in telecommunication is the *broadcasting* of information from a central point to a wide audience, and called commercial broadcasting. Such a system can operate over lines that connect the central point to all the different destinations or receiving points. For example, a public address system enables information to be broadcast to all parts of a factory or similar large organization, as illustrated in Fig. 3.1.

This idea of commercial broadcasting can be extended to a much wider audience and over much larger distances by using a radio transmitter to radiate the information through the atmosphere for detection by a receiver anywhere within the range of the transmitted radio signal power. This is well known of course to everyone these days, with commercial radio and television programmes being an inescapable part of life. The general principles of a commercial broadcasting radio system is illustrated in Fig. 3.2.

We have considered radio waves being propagated through the atmosphere without wires. These radio waves can pass through insulating material, although some energy is lost in the process, but are *reflected* by conducting surfaces. These facts can cause the propagation of radio waves to be irregular and unpredictable in performance. Fading of signals due to the presence of large buildings, etc. is a familiar effect.

Radio waves can also be propagated through a vacuum, and along a transmission line as already considered. In the case of the line, the energy is transmitted through the insulation separating the conductors.

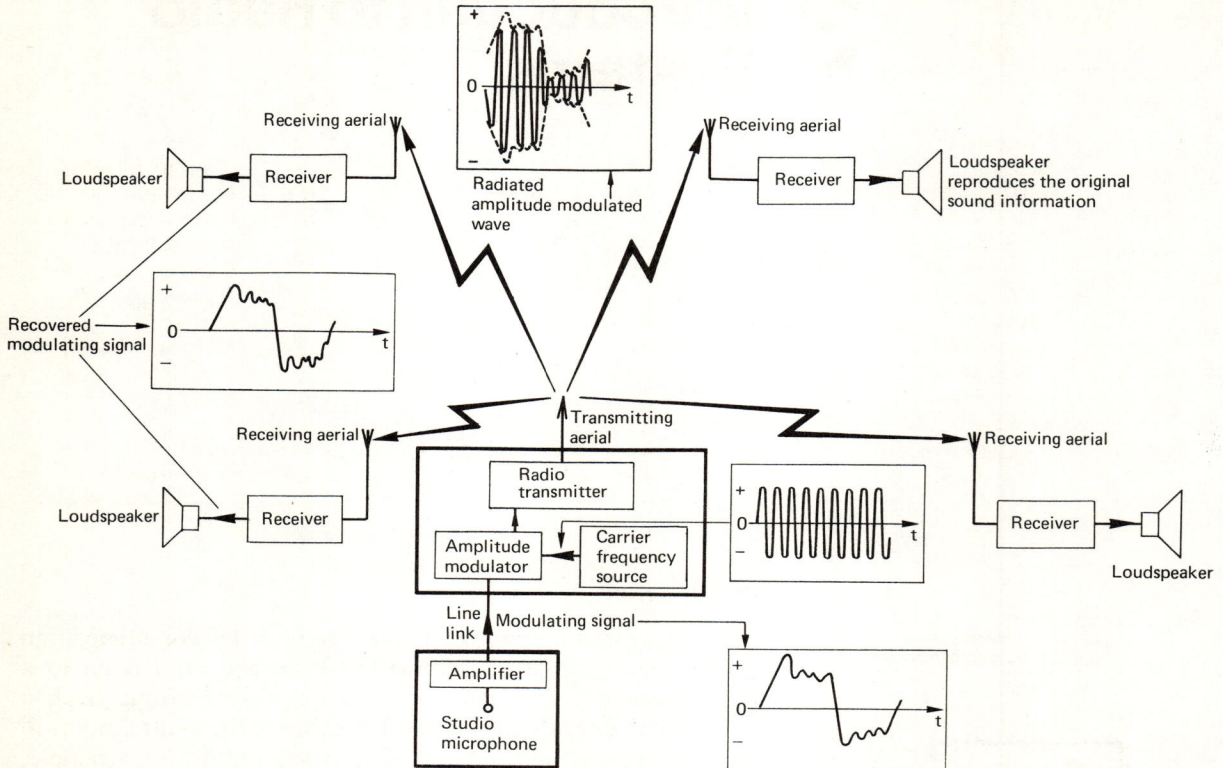

Fig. 3.2 Elements of an amplitude-modulated sound radio broadcasting system

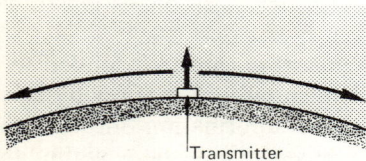

Fig. 3.3a Illustration of ground wave propagation over the earth's surface

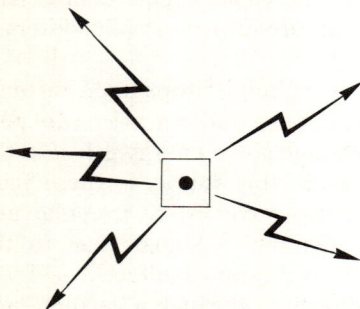

Fig. 3.3b Radio wave broadcasting in all directions by ground wave propagation

Propagation Characteristics of Radio Waves

In Chapter 1 the classification of radio waves was introduced, and each band in the classification had its own particular characteristics.

Low Radio Frequencies (v.l.f., l.f., m.f.)

At these frequencies the radio energy wave leaves the transmitting aerial and follows the curve of the earth's surface in all directions, usually as a GROUND WAVE, as illustrated in Fig. 3.3.

Generally speaking, the distance travelled over the earth's surface depends on the *power* generated by the radio transmitter. The power level of each transmitter is chosen in order to cover a particular broadcasting service area.

As stated in Chapter 1, sound and television radio broadcasting services are examples of *unidirectional* or one-way systems.

Some details of typical long and medium wave radio broadcasting transmitters are given in Table 3.1.

Table 3.1

Programme	Location	Frequency (kHz)	Wavelength (metres)	Waveband	Power (kW)	Programme Coverage
Radio 1	Brookmans Park	1214	247	Medium	50	Greater London
Radio 1	Droitwich	1214	247	Medium	30	Midlands
Radio 1	Newcastle	1214	247	Medium	2	Newcastle area
Radio 2	Droitwich	200	1500	Long	400	Most of UK
Radio 2	Glasgow	1484	202	Medium	2	Glasgow area
Radio 2	Redmoss	1484	202	Medium	2	Aberdeen area
Radio 3	Daventry	647	464	Medium	150	Midlands
Radio 3	Newcastle	647	464	Medium	2	Newcastle area
Radio 3	Swansea	1546	194	Medium	1	Swansea area
Radio 4	Droitwich	1088	276	Medium	150	Midlands
Radio 4	Moorside Edge	692	434	Medium	150	North
Radio 4	Redruth	1457	206	Medium	2	S. Cornwall

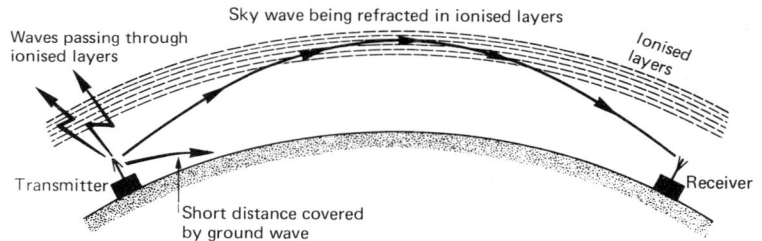

Fig. 3.4 Illustration of skywave propagation via the ionosphere

High Radio Frequencies (h.f.)

At these frequencies the ground wave is absorbed or attenuated very rapidly, but radiation also occurs upwards until the waves reach the IONOSPHERE, which extends approximately from 50 to 400 km above the earths surface. In the ionosphere the gases present are subject to ultraviolet radiation from the sun. The molecules lose some electrons and so become positively charged ions. At certain heights in the ionosphere, recombination between free electrons and positive ions is less likely than in the lower atmosphere, and regions of high ionization exist in the upper atmosphere. Here in these ionized layers, the radio waves are refracted at particular angles in such a way that they are returned to the earth at some distance from the transmitting aerial. This type of radio wave is called a SKY WAVE, and is illustrated in Fig. 3.4.

By using directional aerials at the transmitter, the sky wave can be made to reach a particular destination at a long distance using relatively low power compared to that for a ground wave over the same distance. This method was widely used in the international telephone network for point-to-point communication before long distance ocean cables and artificial space satellites were introduced.

Table 3.2

Station	Radio 2 (MHz)	Radio 3 (MHz	Radio 4 (MHz)	Power (each programme)	Area
Sutton Coldfield	88.3	90.5	92.7	120 kW	Midlands
Ventnor	89.4	91.6	93.8	20 W	Isle of Wight (south coast)
Wrotham	89.1	91.3	93.5	120 kW	S.E. England
Morecambe Bay	90.0	92.2	94.4	4 kW	N.W. Lancashire

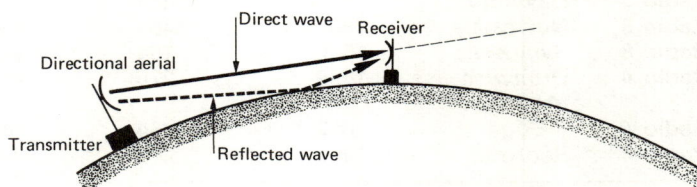

Fig. 3.5 Illustration of line-of-sight propagation by space wave

Very High Frequencies (v.h.f.)

At these frequencies the radio wave energy is propagated through space in straight lines, as is light energy. Using OMNI-DIRECTIONAL aerials (radiating equally in all directions), BBC V.H.F./F.M. sound broadcasting services cover clearly defined areas for Radio 2, 3 and 4 programmes, as indicated in Table 3.2.

By using a DIRECTIONAL aerial, e.g. a parabolic reflector or dish, the energy can be directed towards the horizon, giving a line-of-sight propagation path. This is illustrated in Fig. 3.5.

It should be pointed out that space waves can also be directed down to the ground, where they are reflected, and can thus reach the receiving aerial. This *reflected* wave can cause interference to the *direct* wave because of the longer distance it travels, which means that the reflected wave arrives later than the direct wave.

It should also be fairly clear that the distance over which space waves can be used depends on the heights of the transmitting and receiving aerials. So wherever possible these are mounted on masts or towers erected on high ground.

Space-wave propagation is used in the public telephone network for multi-channel microwave radio relay systems as an alternative to the multi-channel coaxial cable links previously mentioned in Chapter 2.

Fig. 3.6 Two-way radio telephone link using line-of-sight space wave (Typical frequencies in the 2 GHz range. Transmitter power typically 10 W.)

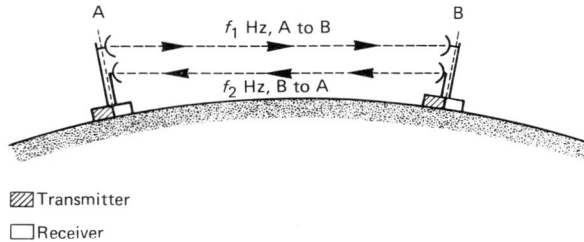

f_1 Hz, A to B
f_2 Hz, B to A

Transmitter
Receiver

f_1 Base to mobiles
f_2 Mobiles to base

Base station

Transmitter/receiver

Fig. 3.7 Two-way mobile radio telephone system
(f_1 typically 164.5 MHz at 15 W
f_2 typically 160.0 MHz at 5 W)

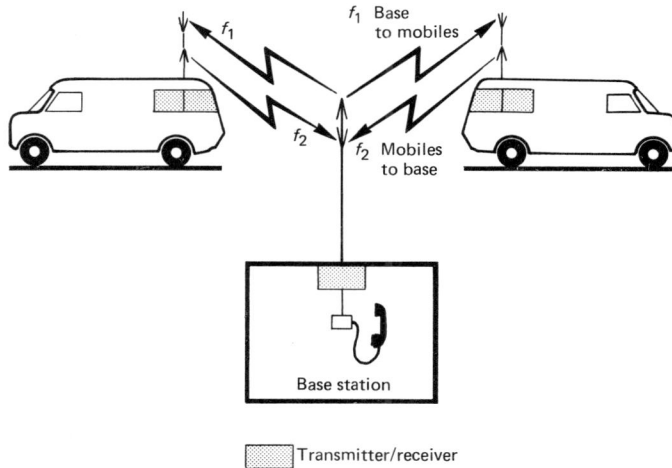

The point-to-point radio systems using sky waves or space waves can be included in a national or international telephone network, but clearly these systems must be capable of transmitting information in *both* directions if a telephone conversation is to be possible. It was shown in Chapter 1 that in order for this to occur there must be a complete telecommunications channel in both directions to form a two-way or bothway circuit. This means that at each end of the system there must be a transmitter and a receiver.

It should be fairly obvious by a little thought that if a large amount of transmitter power is needed to enable a radio wave to reach the distant receiver, then the receiver at each end would be subjected to a very much more powerful signal from its own transmitter than that received from the distant end. It is usual, therefore, to allocate different radio carrier frequencies for the two directions of information transmission. This is illustrated in Fig 3.6. and Fig. 3.7.

Furthermore, with h.f. point-to-point links over long distances, very large transmitter powers may be required (e.g. 30 kW). In this situation it is usual to separate the transmitter and receiver at each end geographically in order to prevent a receiver being swamped by the transmitter at the same end of the system. This is illustrated in Fig. 3.8.

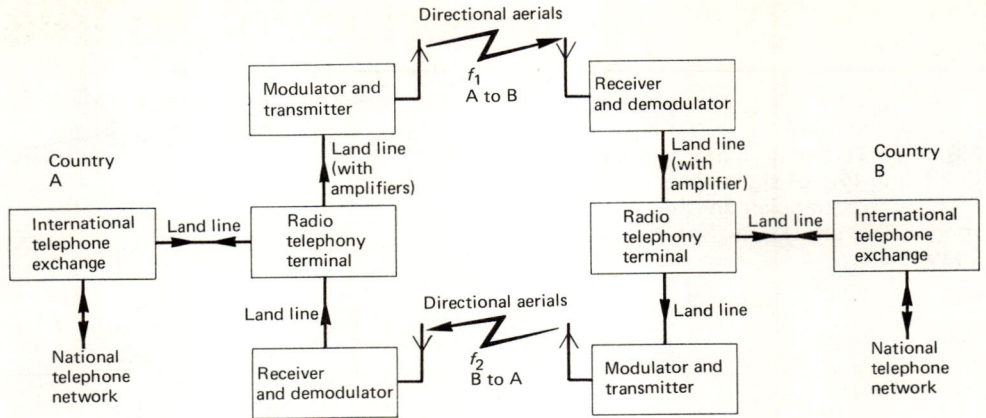

Fig. 3.8 International telephone call routed over two-way high-frequency radio link
(Frequencies in the range 3 to 30 MHz. Typical power 30 kW.)

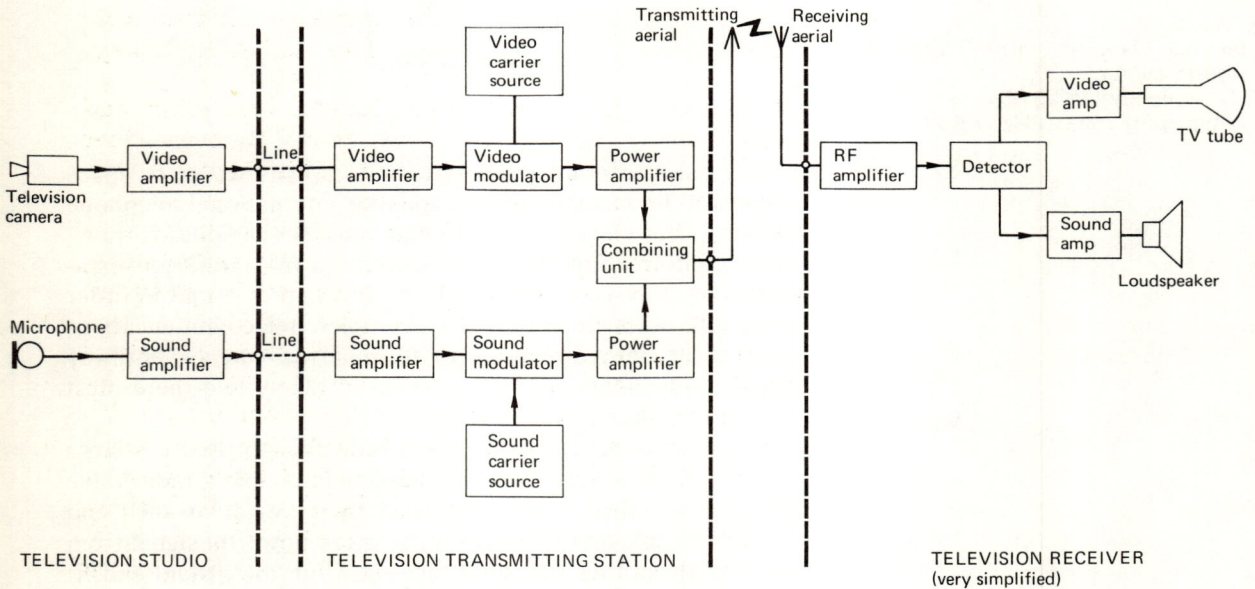

Fig. 4.1 Simple principles of a television broadcast system

4 Introduction to Television

Introduction

In Chapter 1 it was seen that, in any telecommunication system, some form of original information energy is converted by a transducer into an electronic signal to be transmitted to a distant point by a line or radio link, where another transducer converts the electronic signal back into the original energy form.

Television systems can be either monochrome (black and white) or colour, and they are quite different, although colour must be compatible with monochrome. This chapter is concerned only with monochrome systems.

A television system uses one or more television cameras to convert the light energy of a natural moving visible scene, either in a television studio or outdoors, into an electronic signal. Alternatively, the signal may be obtained from a video tape recorder, from telecine machines, or from slide scanners. These last two convert films or photographic slides into appropriate signals. This signal is usually conveyed by line to a television transmitting station where it modulates a carrier source, and the resultant vision-modulated carrier wave is passed to the transmitting aerial to be radiated in all directions as a broadcast vision signal.

At the same time, the sound energy information associated with the visible scene is picked up by a microphone and converted into an electronic signal which is also passed by line to the transmitting station to modulate a separate carrier source. The resultant sound-modulated carrier wave is then passed to the transmitting aerial to be radiated into the atmosphere along with the vision-modulated carrier wave.

Within a certain distance from the TV transmitting aerial, according to the amount of radio-frequency power radiated, a TV receiving aerial can pick up the combined vision and sound

Frequency modulated
sound wave

Amplitude modulated
video wave

TV receiver
aerial

Combined sound
and video modulated
waves

RF
amplifier

Mixer and
IF amp

Detector

Local
oscillator
waveform

Local
oscillator

Sound
IF
amplifier

Sound
detector

Audio
amplifier

Sound signal

Loudspeaker

Frequency
modulated
sound signal

Video
signal

Video
amplifier

Sync
separator

Synchronizing
pulses

Horizontal
time base

EHT
supply

Vertical
time base

Line
deflection
current

Frame
deflection
current

Vision signal

Deflection
coils

Television
tube

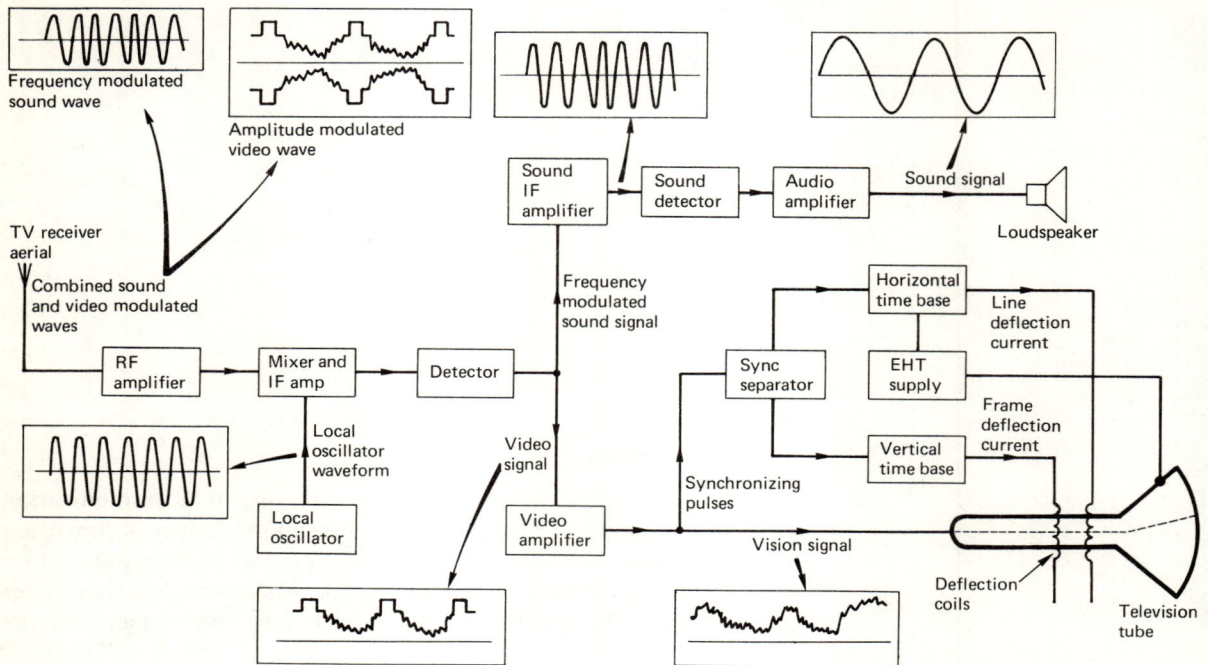

Fig. 4.2 Principles of a television receiver
(using f.m. sound signal and negative-modulation vision signal)

modulated wave to pass it to a TV receiver. The receiver amplifies the received signal, and then separates the vision and sound components after a demodulation process. The demodulated vision signal is passed to a cathode ray tube to reproduce as closely as possible the original visible moving scene at the transmitting end. The demodulated sound signal is passed to a loudspeaker to reproduce as closely as possible the original sound information associated with the visible scene.

The simple principles of a television broadcast system are illustrated in Fig. 4.1 and the basic principles of the TV receiver arrangement are shown in Fig. 4.2.

Principle of Moving Pictures

The reader will perhaps be familiar with the production of moving pictures by a cine-film projector. A number of "still" pictures are presented on a screen to the human eye in rapid succession, each "still" picture being slightly different from the preceding one. The human eye has a characteristic called

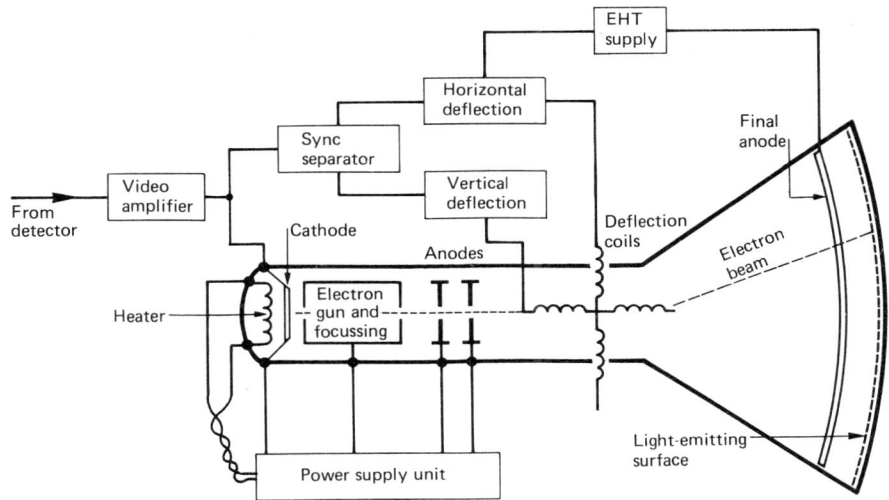

Fig. 4.3 Simple principles of cathode ray tube for television reception

"persistence of vision," by which the signal to the brain caused by a light source reaching the eye survives for a very short time after the light source is removed. If "still" pictures are presented one after another to the human eye at a rate of more than 16 per second, an illusion of a moving scene is created without any significant flicker. A television system must therefore be designed to present pictures to the human eye from the TV receiver at a rate of 16 per second or more.

Principle of the Cathode Ray Tube

It was stated in the introduction that the conversion of the electronic vision signal back into light energy is achieved by a cathode ray tube, the simple principles of which are illustrated in Fig. 4.3.

The tube consists of an evacuated glass envelope having a narrow cylindrical end which flares out from its "neck" into a wider rectangular face forming a viewing screen. A cathode is placed in the end of the cylindrical tube, and is heated to emit electrons.

An arrangement called an "electron gun" is associated with the heated cathode and this gun serves to focus the emitted electrons into a narrow beam which is fired along the tube under the influence of the positive potential applied to an arrangement of anodes. The electron beam can be moved in horizontal and vertical directions by magnetic fields produced by currents passing through deflection coils clamped around the outside of the "neck" of the tube.

The inside surface of the rectangular viewing screen is coated with a light-emitting material. If the electron beam fired along the tube strikes the screen coating with sufficient velocity, the energy of the electron beam causes light to be emitted from the surface coating, and a small spot of light is seen on the tube screen when viewed from the front.

Principle of Scanning

By passing suitable electric currents through the deflection coils, magnetic fields are produced which can control the path of the electron beam along the tube by simultaneous horizontal and vertical forces, and so the small spot of light can be moved around the screen at will. (See Fig. 4.4.)

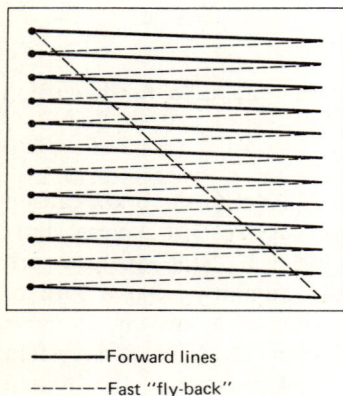

To produce a picture, the small spot of light is positioned initially in the top left-hand corner of the rectangular screen as viewed from the front. It is then moved rapidly across the screen by the horizontal deflection force. When the end of the first line is reached, the spot is returned very rapidly to the left-hand side of the screen but positioned slightly below the starting point of the first line. This very rapid return is called the FLY-BACK of the spot of light. A second line is now traced out by the horizontal deflection force, and again the fly-back is carried out. The positioning of the spot at the beginning of each line just below the previous line is achieved by the vertical deflecting force. This process is repeated until the spot of light reaches the bottom right-hand corner of the rectangular screen, and one complete picture has been traced out or scanned by the spot of light in successive horizontal lines.

The spot is now returned to the top left-hand corner of the screen, and a second picture is traced out or scanned in the same way as the first one. If this is repeated rapidly so that more than 16 pictures per second are traced out, and if the intensity of the spot of light is constant, the moving spot appears as a complete white picture or "raster".

The vision signal, demodulated from the received television signal, is now applied to the cathode ray tube to control the intensity of the electron beam passing along the tube. The amount of light energy emitted by the screen material will vary in accordance with the intensity of the electron beam and so the light energy output from the TV screen will be a reproduction of the light energy picked up by the TV camera or other equipment, and the illusion of a moving scene is presented to the observer. This picture appears in black, white and all intermediate shades of grey, and is called a monochrome picture.

———— Forward lines

------- Fast "fly-back"

Fig. 4.4 Simple principle of scanning a picture

Number of Lines

The number of lines used in television systems has varied in different countries over the years. For example, 405, 525, 625 and 819 lines have been used.

In the UK the original BBC and ITA channels in the v.h.f. Band I and Band III (ranging approximately from 30–300 MHz) used 405 lines, but more recently new TV transmitting stations have been installed using 625 lines for BBC and ITA channels in the u.h.f. Band IV and Band V (ranging approximately from 300–3000 MHz).

The 405-line system uses amplitude-modulation for both vision and sound channels, but the 625-line system uses amplitude-modulation for the vision channel and frequency-modulation for the sound channel.

Aspect Ratio

The shape of the rectangular picture as viewed from the front of the TV receiver tube is defined by the ratio of the picture width to picture height. This is called the VISUAL ASPECT RATIO, and is different from the electrical aspect ratio in terms of the number of lines because not all the lines are used to convey actual picture information, some are needed to transmit synchronization signals as will be explained later in this chapter. The visual aspect ratio used in both systems in the UK is 4:3.

Interlaced Scanning

In addition to at least 16 pictures per second being necessary to create the illusion of a moving picture, it has also been found that the number of pictures per second must be the same as the frequency of the a.c. mains supply in order to avoid "hum bars" appearing across the screen. So in the UK and Europe, 50 pictures per second are needed, and in the USA 60 pictures per second.

Using this number of pictures per second with the simple principle of scanning results in a vision electronic signal having a very large frequency bandwidth, and this would mean that a limited number of TV transmitting stations could be accommodated in the allocated frequency bands. So it may not be possible to provide complete coverage of a particular country.

In order to reduce the bandwidth of the vision signal and therefore allow more TV transmitters to be used, a technique called INTERLACED SCANNING has been devised. Each complete picture is divided into two frames or fields which are scanned and transmitted one after the other, and then reas-

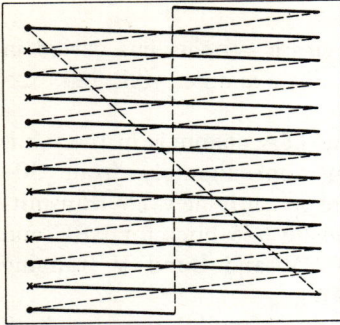

Fig. 4.5 Principle of interlaced scanning of alternate lines

sembled at the TV receiver. So although 50 frames or fields are transmitted every second to avoid mains "hum bars" (in the UK), only 25 complete pictures are transmitted every second. The vision signal therefore contains only half the information compared with 50 pictures per second, and so the bandwidth is also reduced by a half. This allows twice as many TV transmitters to be allocated in the available bands.

The interlaced scanning of the two frames or fields making up each complete picture is achieved by scanning *alternate* lines by the spot on the TV tube screen. After one field of alternate lines has been scanned, the spot returns to fill in the gaps between these lines as it scans the second field. This principle is illustrated simply in Fig. 4.5.

Producing the Vision Signal

We have seen that the variations of the vision signal level or strength control the intensity of the electron beam in the TV receiver tube to produce the appropriate amount of light from the TV screen. At the other end of the TV system, the TV camera produces this vision signal by using the same interlaced scanning principle as described for the TV tube.

Very simply, the visible scene to be transmitted is focussed by the optical lens system of the TV camera on to a light-sensitive surface which absorbs light energy according to the instantaneous scene being focussed by the TV camera.

An electron beam scans the light-sensitive surface and the strength of the electron beam is varied by the amount of light energy absorbed by each small spot on the light-sensitive surface. This variation of the electron beam strength is converted into a varying voltage that constitutes the vision signal containing a range of frequencies which is much wider than that for speech or music. This will be explained in greater detail later in the course, but as a guide it can be stated that the *bandwidths* of vision signals in the UK systems are as follows:

    For the 405-line system   0 to 3 MHz
    For the 625-line system   0 to 5.5 MHz

Synchronizing the TV Camera and the TV Receiver Tube

From the simple principle of interlaced scanning it should be fairly obvious that the electron beam scanning the screen of the TV receiver tube must be in exactly the same position at all times as the electron beam that is scanning in the TV camera. This is achieved by including SYCHRONIZING PULSES along with the vision information signal. Both LINE

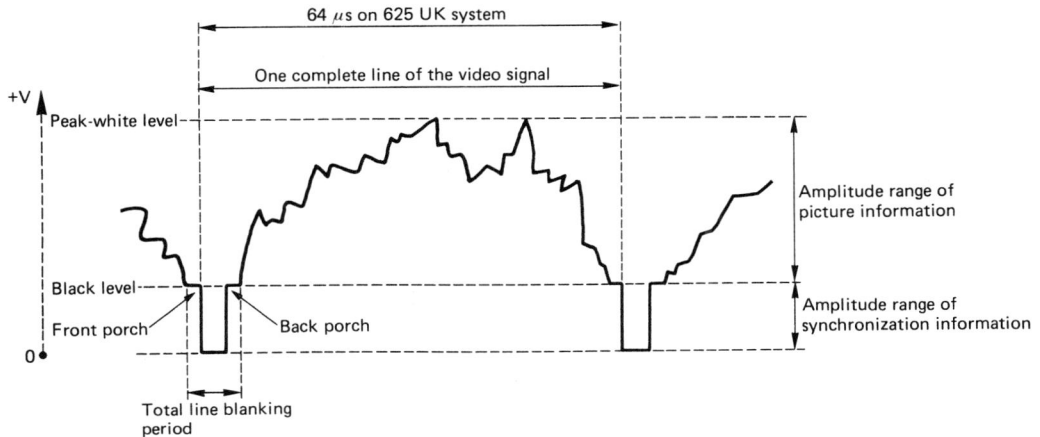

Fig. 4.6 Simple principle of video signal

and FRAME (or field) synchronizing pulses are used to ensure that the correct line of the appropriate frame is being produced by the TV receiver tube. These synchronizing pulses are separated from the vision signal in the TV receiver, and are used to trigger line and frame time-base circuits which supply the currents for the deflection coils to position the spot of light on the TV receiver screen. This is illustrated in Fig. 4.2 given earlier.

The Video Signal

The combination of vision (or picture) signal and synchronizing pulses is called a VIDEO SIGNAL.

Also included is a "blanking period" or picture suppression period, to allow time for the fly-back of the spot from one line to the next and from the end of one field or frame to the beginning of the next. This is illustrated very simply in Fig. 4.6, where it is seen that the picture and synchronization information are separated by time and amplitude.

Fig. 4.6 illustrates what is known as a *positive-going* (or positive modulation) video signal. It should be pointed out that some TV systems use a *negative-going* (or negative modulation) video signal which is upside-down compared with Fig. 4.6.

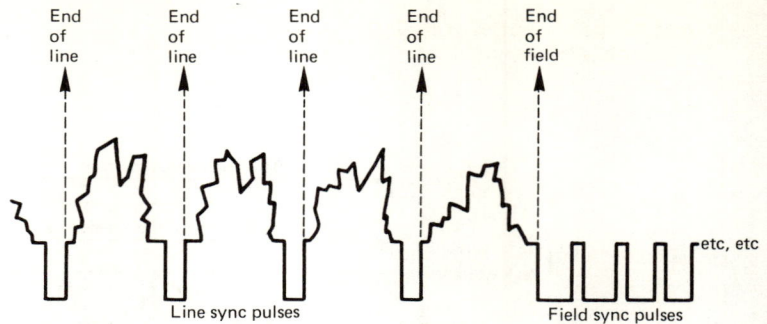

Fig. 4.7 Simple illustration of line and field sync-pulses

The line and field (or frame) synchronization pulses are included in the blanking level amplitude region. *Line* synchronization pulses are simple narrow pulses, whilst *field* synchronization pulses are a series of broader pulses. This is illustrated simply in Fig. 4.7.

The field synchronization pulses take a time equivalent to a number of lines, according to the system being used, and are followed by a number of suppressed lines.

In the *UK 405-line system*, the field sync pulses occupy the equivalent of 4 lines, followed by 10 suppressed lines, giving an overall suppression equivalent to 14 lines per field. So, for each complete picture consisting of 2 successive interlaced fields, 28 lines are used for field synchronization giving $(405 - 28) = 377$ lines containing the picture or vision information.

In the *UK 625-line monochrome system*, 20 lines are used for each field synchronization, giving $(625 - 40) = 585$ lines containing the picture or vision information.

Bandwidth of Video-modulated Carrier Signals

It was seen in Chapter 2 that when a carrier is amplitude-modulated by an information signal, the bandwidth of the modulated wave is *double* the highest modulating signal frequency. It was also stated, earlier in this chapter, that a vision signal has a wide range of frequencies, with a highest frequency of approximately 3 MHz for a 405-line system and 5.5 MHz for a 625-line system. Using normal double-sideband amplitude modulation would therefore require a bandwidth of approx 6 MHz for 405 lines and 11 MHz for 625 lines, *plus* a bandwidth for the separate sound-modulated carrier waveform.

To reduce the bandwidth needed for each transmission and therefore allow more transmissions in a given frequency band, the modulated wave is passed through a VESTIGIAL SIDEBAND FILTER, which suppresses part of *one* of the sidebands. This is illustrated simply in Fig. 4.8.

Fig. 4.8 Bandwidths of typical TV systems in UK

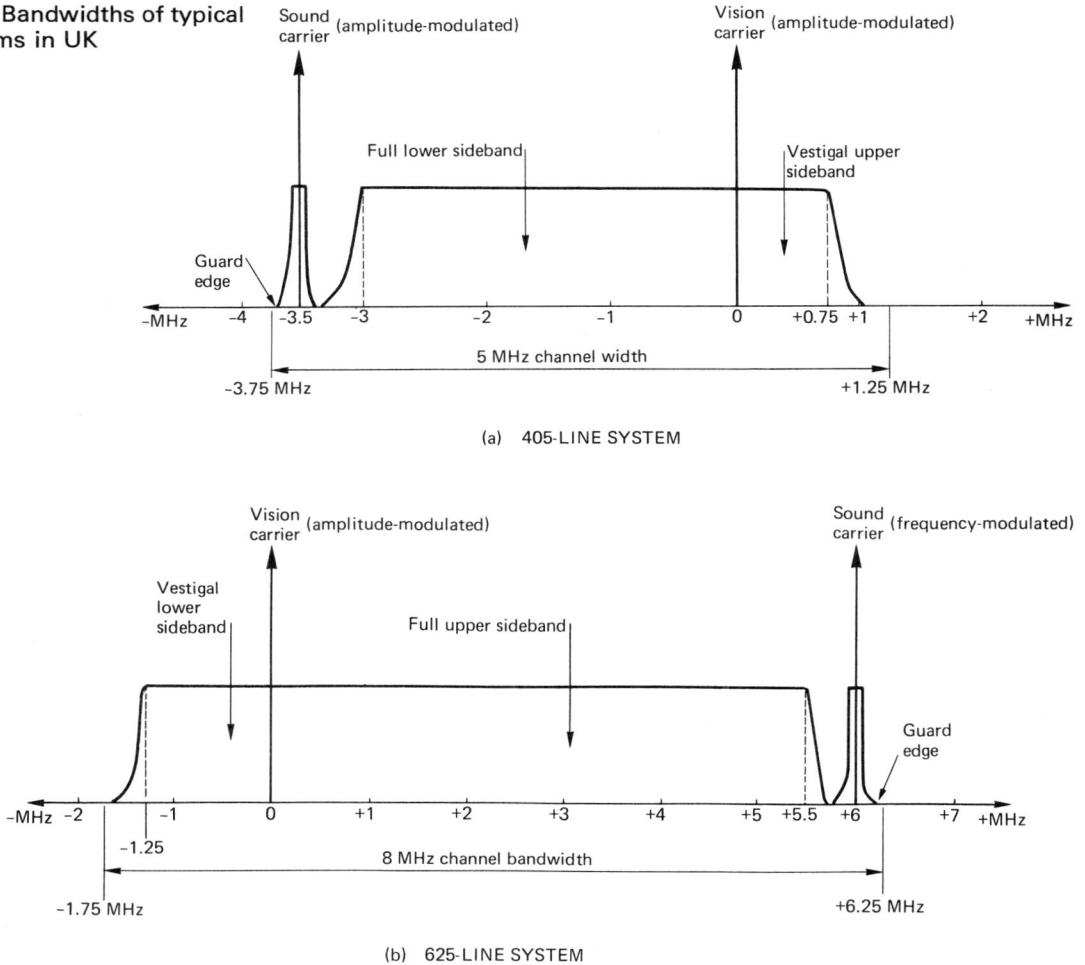

Sound carrier (amplitude-modulated)

Vision carrier (amplitude-modulated)

Full lower sideband

Vestigal upper sideband

Guard edge

−MHz −4 −3.5 −3 −2 −1 0 +0.75 +1 +2 +MHz

5 MHz channel width

−3.75 MHz   +1.25 MHz

(a)   405-LINE SYSTEM

Vision carrier (amplitude-modulated)

Sound carrier (frequency-modulated)

Vestigal lower sideband

Full upper sideband

Guard edge

−MHz −2 −1 0 +1 +2 +3 +4 +5 +5.5 +6 +7 +MHz

−1.25

8 MHz channel bandwidth

−1.75 MHz   +6.25 MHz

(b)   625-LINE SYSTEM

In Fig. 4.8, frequencies are shown relative to the allocated vision carrier frequency. It will be seen that in a 405-line system the sound carrier is 3.5 MHz *below* the associated vision carrier, and the *upper* sideband is restricted. By comparison, in a 625-line system, the sound carrier is 6 MHz *above* the associated vision carrier, and the *lower* sideband is restricted. Notice also in Fig. 4.8 that there is a "guard edge" between the sound information bandwidth and the extreme edge of the particular channel.

In the UK, vision and sound carrier frequencies are allocated to the various channels in the TV bands in such a way that each complete channel slots into a particular section of the band so that interference is avoided between adjacent channels. Typical TV channel frequencies are given in Table 4.1.

Table 4.1

Band 1 (405-lines) (40-68 MHz)	Carrier Frequencies (MHz)		Band IV (625-lines) (470–610 MHz)	Carrier Frequencies (MHz)		Band V (625-lines) (610–940 MHz)	Carrier Frequencies (MHz)	
Channel	Sound	Vision	Channel	Sound	Vision	Channel	Sound	Vision
1*	41.5	45.0	21	477.25	471.25	39	621.25	615.25
2	48.25	51.75	22	485.25	479.25	40	629.25	623.25
3	53.25	56.75	23	493.25	487.25	.	.	.
4	58.25	61.75	24	501.25	495.25	.	.	.
5	63.25	66.75	25	509.25	503.25	68	etc.	etc.
			.	.	.			
			.	.	.			
			38	etc.	etc.			

* originally double sideband

5 Principles of the Telephone Instrument

Introduction

In Chapter 1, the origin of the word "telephone" was seen to be "speaking at a distance", and it was also seen that a transducer is required to change the sound energy generated by the speaker's voice into an electronic signal. Then, in Chapter 2 it was seen that for national and international telephone systems the important range of speech frequencies of the human voice is 300–3400 Hz.

It follows therefore that telephone instruments and the lines connecting them together must be capable of handling alternating currents in this frequency range. Also in Chapter 1, Fig. 1.1, a one-way or *unidirectional* telecommunication channel was illustrated, with an originating transducer at the sending end and a reproducing transducer at the receiving end. It was stated that to allow *two-way* transmission of information signals it is necessary to duplicate the arrangement of Fig. 1.1 in the opposite direction. This implies that other suitable transducers are required at each end, connected by another line.

The simple outline of such a bothway telephone circuit is illustrated in Fig. 5.1. The nature of the interconnecting lines indicated in Fig. 5.1 will be considered in the next chapter.

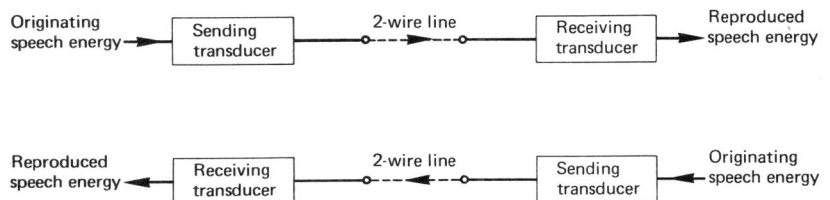

Fig. 5.1 Simple principle of two-way telephone circuit, using two lines

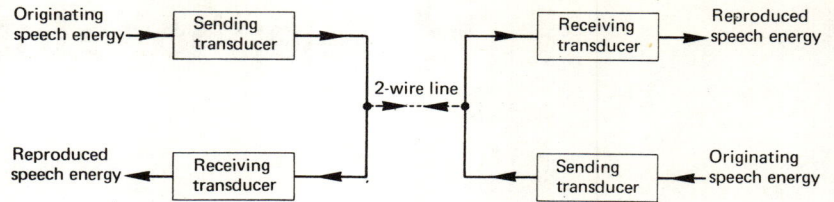

Fig. 5.2 Simple principle of two-way telephone circuit, using a single interconnecting line

For a number of reasons, one of them being economy in the provision of these lines, it was decided to produce a telephone system in which the electronic speech information signals in both directions are carried by a *single* line, as illustrated in Fig. 5.2.

The decision to use this arrangement has a very important influence in the design of a suitable telephone instrument, and in fact was the direct cause of perhaps the most difficult problem encountered, the one of SIDETONE (which will be considered in detail in the second year of this course). Briefly, the problem arises from the fact that speech energy signals generated by a sending transducer are passed to the associated receiving transducer as well as being passed to line.

The Sending Transducer

The reader will no doubt be familiar with the use of the MICROPHONE as a transducer which picks up sound energy waves and converts them into electronic signals. This word "microphone" is widely used in public address systems, radio and television broadcasting, tape recorders, and so on. But as the telephone system developed, the term *telephone transmitter* came to be accepted for general use.

There are several types of microphone now in use for the systems mentioned previously, but the one chosen as a standard for the telephone is the CARBON GRANULE transmitter.

The Carbon Granule Transmitter

To understand the simple principles of the transmitter it is necessary to consider the nature of the sound energy waves produced by the human voice. Normally, at ground level, the atmosphere can be considered as columns of air having a normal pressure of approximately 1.05 kilogrammes per square centimetre (15 pounds per square inch). Any source of

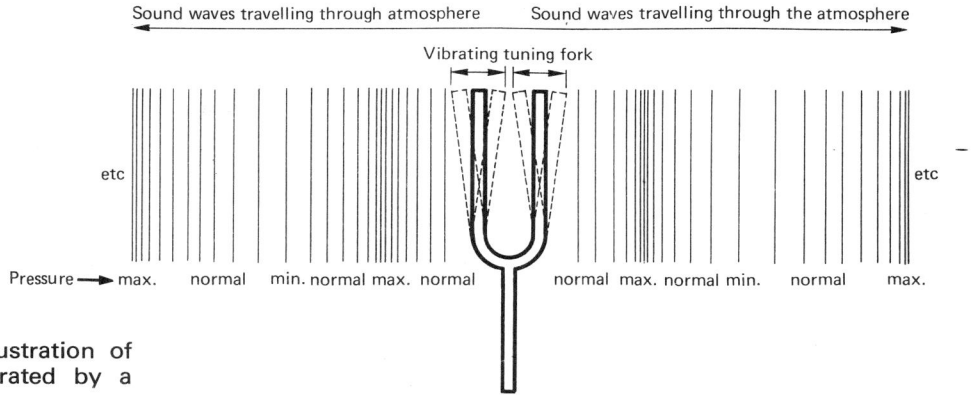

Fig. 5.3 Simple illustration of sound waves generated by a tuning fork

sound energy has some element which vibrates and causes variations of atmospheric pressure above and below the normal value, and these variations are passed through the atmosphere as sound energy waves gradually decreasing in value until the energy is used up.

One example of a simple source of sound is a tuning fork, which vibrates at a particular audio frequency according to its physical size. The way in which sound energy waves can be represented is illustrated in Fig. 5.3.

When a person speaks, the vocal chords set up vibrations of the columns of air which produce a speech sound information signal. These vibrations reach the telephone transmitter where a *diaphragm* responds to the vibrations and begins to vibrate itself. Increases in pressure move the diaphragm inwards, and decreases in pressure allow the diaphragm to move outwards. The simple principle of this diaphragm is illustrated in Fig. 5.4.

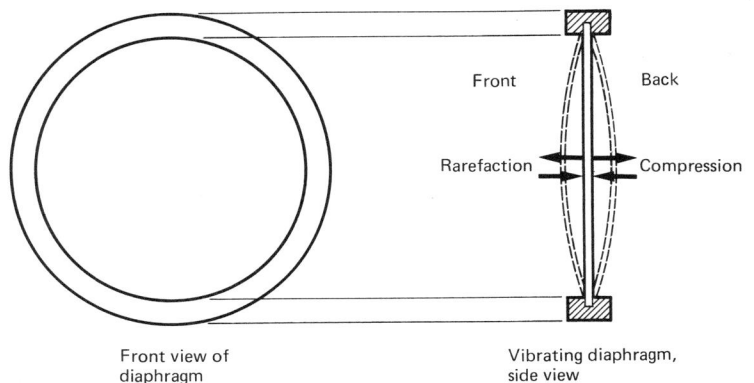

Fig. 5.4 Simple principle of telephone transmitter diaphragm

The vibrations of the diaphragm are now used to produce a varying electric current that forms an electronic information signal that is ideally the direct copy or analogue of the speech information energy.

It is necessary now to recall briefly the relationship between voltage, current and resistance in an electrical circuit. If a source of electrical energy (e.g. a battery), which has an electromotive force (e.m.f.) of E volts and zero internal resistance, is connected to a circuit which has a resistance (opposition to current flow) of R ohms, then the value of electric current, I amperes, flowing in the circuit is given by

$$\text{Current flowing} = \frac{\text{Electromotive force}}{\text{Resistance}}$$

Using standard symbols this is written as $\qquad I = \dfrac{E}{R}$

The circuit diagram for this is given in Fig. 5.5.

Therefore, for a given value of e.m.f., the current will increase if the resistance is reduced, and the current will decrease if the resistance is increased.

Now, if it can be arranged for the vibrations of the transmitter diaphragm to vary the resistance of an electric circuit, then the current in the circuit will vary in sympathy with the diaphragm as it vibrates due to the speech sound energy waves. This is achieved by attaching a carbon block or electrode to the diaphragm, and placing this electrode inside a chamber containing hard polished carbon granules. Another carbon block or electrode is fixed inside the chamber. The simple principle is illustrated in Fig. 5.6, and the simple equivalent circuit is shown in Fig. 5.7 where the carbon granule chamber and diaphragm are represented by a variable resistance, R ohms.

The form of the varying electric current produced by the diaphragm vibrating under sound energy waves is given in Fig. 5.8.

The variation of resistance is due to the fact that the varying pressure on the hard polished carbon granules produces different areas of contact between adjacent granules, as illustrated in Fig. 5.9. The battery is therefore essential in order to provide the direct current flowing through the carbon granule transmitter, otherwise the transmitter cannot function. This direct current is called the *polarizing current*. Some types of transmitter or microphone do *not* require this polarizing current, but the carbon granule transmitter does.

One way of passing the electronic speech information signal to the receiving transducer of the distant telephone is shown in Fig. 5.10. (This diagram uses a pictorial representation of a telephone transmitter, not the standard symbol.) In this ar-

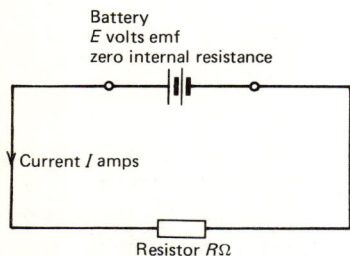

Fig. 5.5 Illustration of simple electric circuit

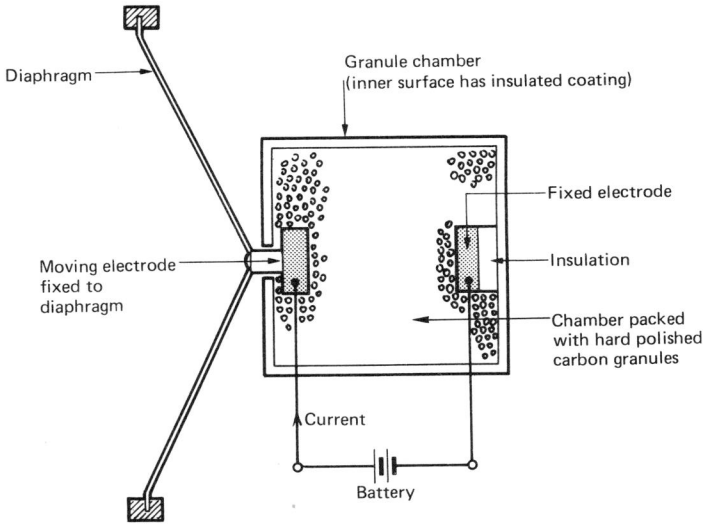

Fig. 5.6 Simple principle of carbon granule telephone transmitter

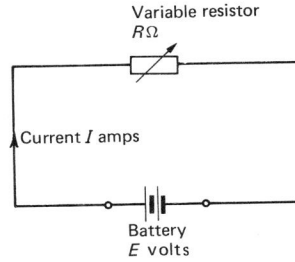

Fig. 5.7 Simple electric circuit representing carbon granule telephone transmitter

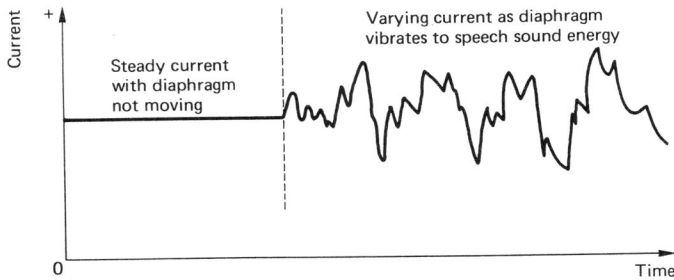

Fig. 5.8 Varying d.c. produced by carbon granule transmitter

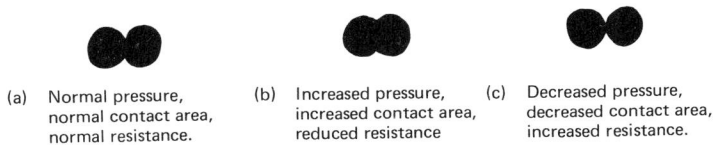

(a) Normal pressure, normal contact area, normal resistance.

(b) Increased pressure, increased contact area, reduced resistance

(c) Decreased pressure, decreased contact area, increased resistance.

Fig. 5.9 Illustration of varying resistance of carbon granule telephone transmitter

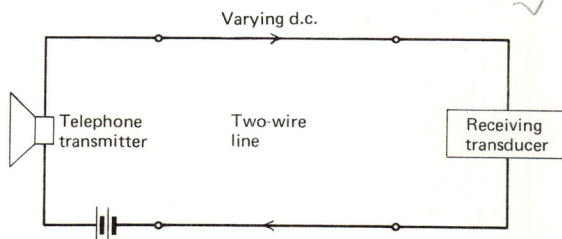

Fig. 5.10 Simple one-way telephone circuit

rangement the resistance of the carbon granule transmitter is connected in series with the resistance of the line, which consists of two wires insulated from each other and from earth.

If the line is long, the line resistance may be much greater than the transmitter resistance, and so the *variations* of the transmitter resistance as the diaphragm vibrates will be very small compared with the total circuit resistance. The variations of current will also be very small and the receiving transducer at the distant telephone will not be able to respond satisfactorily.

This difficulty can be overcome by using a battery of much higher e.m.f., or by isolating the transmitter resistance from the line resistance by means of a transformer, as shown in Fig. 5.11 (which uses the standard symbol for a telephone transmitter). The action of the transformer is such that, with a steady current flowing in the primary winding, no current flows in the secondary winding connected to line. But when the transmitter current varies, an e.m.f. is induced into the secondary winding, by mutual inductance, to drive a current in the line. An increase in primary current produces an induced e.m.f. in the secondary with a certain polarity, and a decrease in primary current reverses the polarity of the induced e.m.f.

This results in an *alternating* electronic information signal current flowing in the line and the receiving transducer at the distant transducer. This a.c. signal contains frequencies in the range 300–3400 Hz, as explained previously.

It should be pointed out that this arrangement requires a battery of low e.m.f. (e.g. 3V) at each telephone instrument. At least this was the case with early telephones used in public systems, but in the second year of the course it will be shown that modern telephones do not require this "local battery" at the telephone subscriber's premises. It should also be pointed out that the transformer shown in Fig. 5.11 is, by accepted practice in telephone language, called an "induction coil".

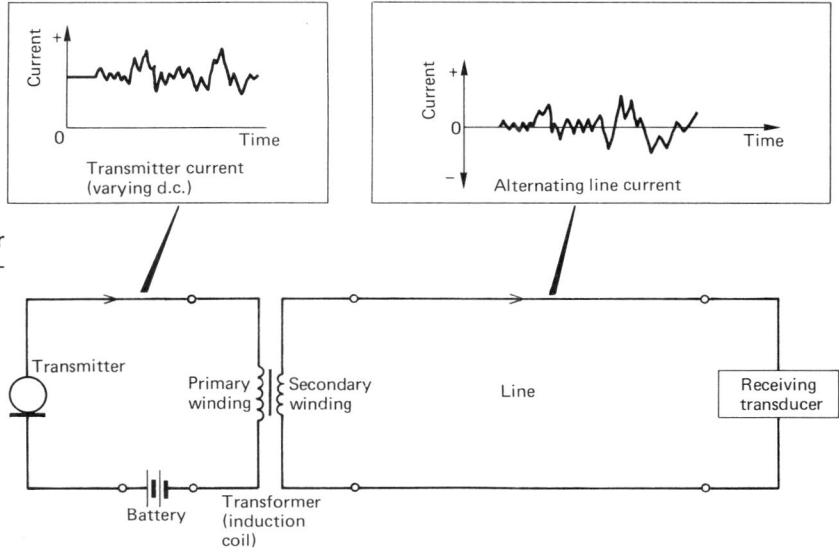

Fig. 5.11 Use of transformer (induction coil) in simple one-way telephone circuit

The Receiving Transducer

Several types of transducer have been used over the years for reproducing the speech sound information energy from the alternating electronic speech information signal. The type used in modern telephone instruments is called the ROCKING ARMATURE RECEIVER.

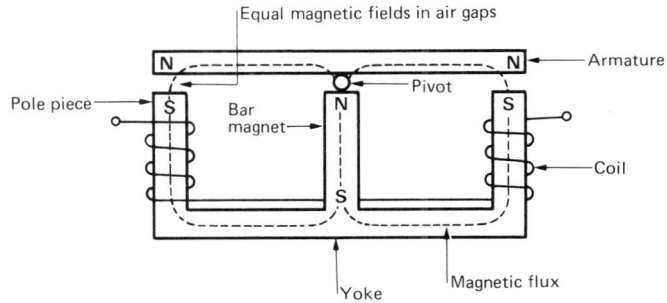

Fig. 5.12 Simple principle of rocking armature receiver

Principles of the Rocking Armature Receiver

It consists essentially of a permanent bar magnet with extended soft iron yoke and pole pieces, as shown in Fig. 5.12. Coils of insulated wire are wound around the pole pieces, and these coils are connected in series with the two-wire line from the distant telephone. An armature is pivoted at its centre and arranged so that it is held horizontal by the permanent magnetic field as long as no current is flowing from line into the coils, which is the situation when the magnetic fields in the gaps between the pole pieces and the armature are equal.

When a current flows from the line through the receiver coils, electromagnetic fields are produced by the coils such that the field in one gap is increased and the field in the other gap is decreased. This causes the armature to be attracted by the strongest field, as shown in Fig. 5.13.

If the current flows through the coils in the opposite direction, the magnetic field strengths in the gaps are reversed, and the armature is attracted by the opposite pole face.

It was shown in Fig. 5.11 that the action of the induction coil at the sending end produces an alternating electronic speech current that contains frequencies in the range 300–3400 Hz. So the current passing through the receiver coils is constantly reversing direction, causing the armature to rock on its pivot from one pole face to another in sympathy with the alternating speech currents from the line. The movements of the armature are transmitted to a diaphragm by a driving pin, as illustrated in Fig. 5.14, and the diaphragm vibrates to generate a sound energy wave that is a reasonable reproduction of the original sound energy information generated by the person talking at the other end of the line.

Simple Local-Battery Telephone Circuit

A carbon granule transmitter and a rocking armature receiver can be combined into a simple telephone instrument, and two such instruments are shown connected together by a two-wire line in Fig. 5.15.

It will be seen from Fig. 5.15 that if a person is speaking at the left-hand telephone, the alternating speech signal in the line flows through *both* receivers in series, so the talker will hear his own voice. As previously mentioned, this is called sidetone, and has a number of practical disadvantages. In a modern telephone instrument the design is such that the level of sidetone is reduced to an acceptable minimum.

Remember also that the design dispenses with a local battery, but the polarizing current for the carbon granule transmitter is obtained from a central battery located in the telephone exchange to which the telephone instrument is connected.

(These two important points will be dealt with in the second year of the course.)

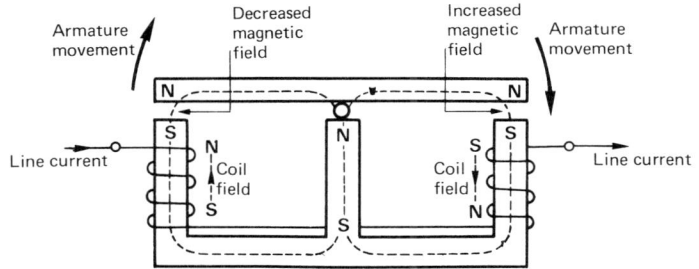

Fig. 5.13 Principle of operation of rocking armature receiver by line current

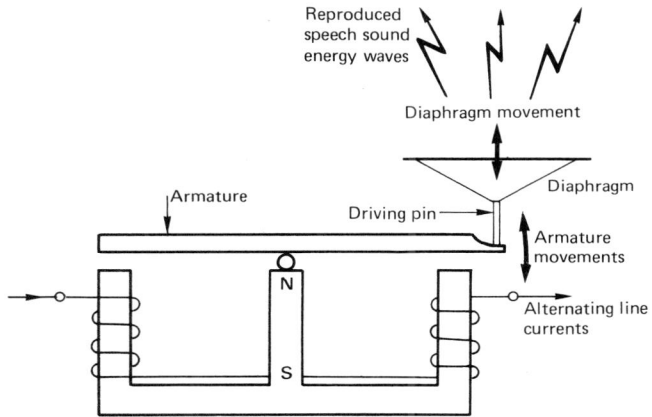

Fig. 5.14 Reproduction of sound energy by rocking armature receiver

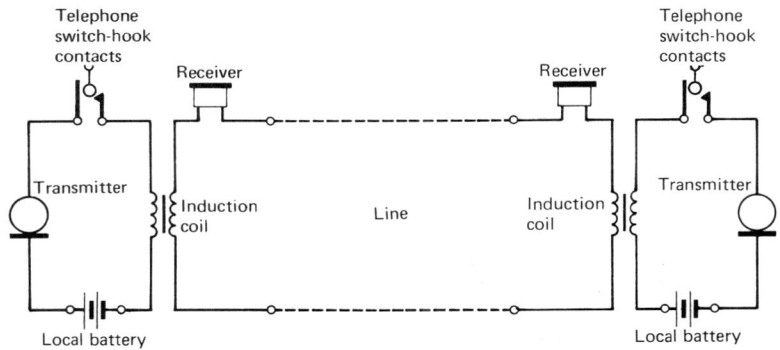

Fig. 5.15 Simple local battery telephone circuit

6 Introduction to Lines, Losses, and Noise

In Chapter 1 the idea of using wires as a line link to convey electronic information signals between two points was introduced as an alternative to using a radio link.

It was also seen in Chapters 3 and 4 that, in radio systems, lines are used to carry information signals from radio or TV studio or radio-telephony terminal to a radio transmitting station.

Then, in Chapter 5, the idea of using a 2-wire line to connect two telephone instruments together was introduced. Interconnecting lines such as these will now be considered very briefly, with more details given in the second year of the course.

Generally, a *transmission line* can be considered as a conductor, or group of conductors, with suitable insulating materials, whose function is to carry electronic information signals. The line can take various physical forms according to the type of information to be transmitted and the distance involved.

Earth Return Circuits

Early morse code telegraph circuits used a *single* conductor or wire to connect two places together. This is illustrated very simply in Fig. 6.1.

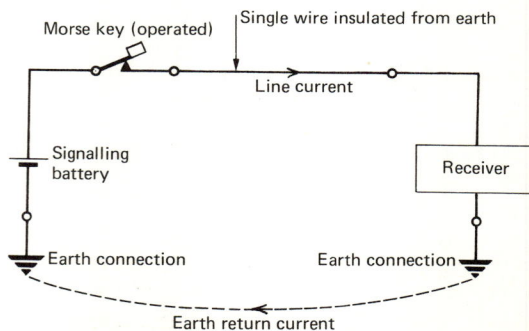

Fig. 6.1 Single wire and earth return telegraph circuit

The earth contains large amounts of different metals and can be used as a return conductor provided that a good connection with low resistance can be made with it. The main disadvantages of this arrangement, apart from the problem of making a good electrical connection to the earth, are

(a) The resistance (or opposition to current flow) of the insulated single wire is greater than the return path through the earth, so the line is unbalanced.

(b) If other circuits also use the same arrangement, the earth is carrying return currents of all the different circuits, and mutual interference between the various circuits can occur.

(c) Power supply circuits which themselves do not carry information signals can also produce interference to earth-return circuits.

Two-wire Lines

The disadvantages of the earth-return system can be largely overcome by using two identical conductors insulated from each other and from earth. The two conductors will now have the same resistance, and are not used by any other circuit.

The simplest form of two-wire line is produced by using bare conductors suspended on insulators at the top of poles. This is illustrated in Fig. 6.2.

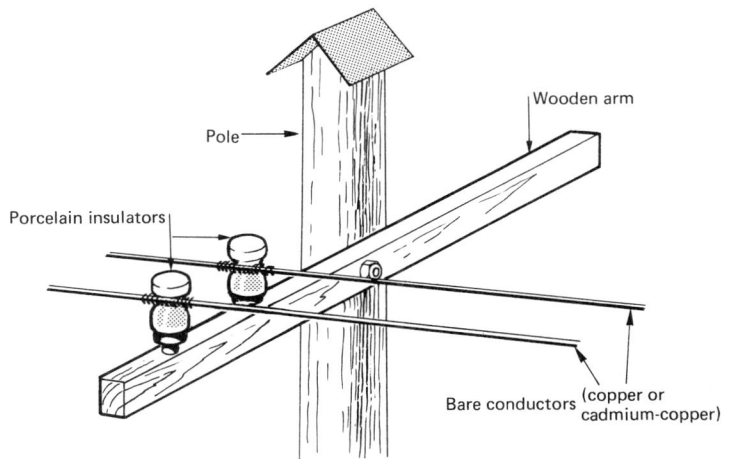

Fig. 6.2 Simple overhead two-wire line

Another type of two-wire line consists of conductors insulated from each other in a cable which also has an outer cover of insulation, as illustrated in Fig. 6.3. Often the two insulated conductors in the cable shown are twisted together along the length of the cable, and are called a *pair*.

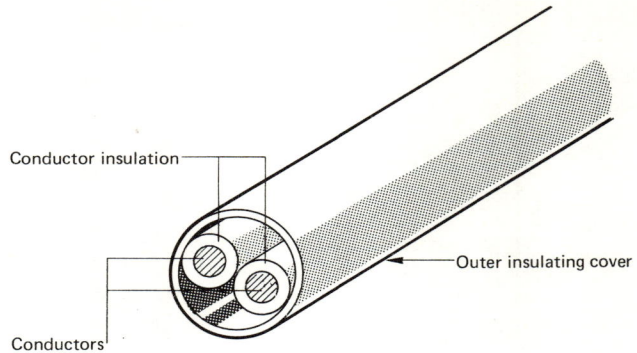

Conductor insulation

Outer insulating cover

Conductors

Fig. 6.3 Simple two-wire cable

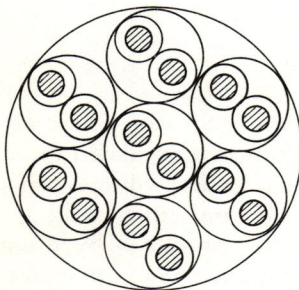

(a)   TWIN-TYPE

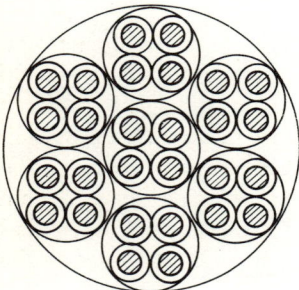

(b)   QUAD-TYPE

Fig. 6.4 Simple illustration of multi-pair cables

Multi-pair Cables

It is often necessary to provide a number of two-wire lines between the same two places, and this is done most conveniently by making a cable with a number of pairs of insulated wires inside it. Sometimes the wires are twisted together in pairs as illustrated in Fig. 6.4a, but sometimes they are provided in fours, or quads, as shown in Fig. 6.4b.

In order to identify the various wires, each one has a colouring on the insulating material around it in accordance with a standard colour code. (This will be dealt with in detail in the second year of the course.)

Coaxial Cables

As the frequency of an alternating current is increased, the current tends to flow along the outer part of a conductor having a circular cross-section. This means that the centre part of the conductor is not carrying current and can be removed. The empty space can then be used for a second conductor, provided it is insulated from the outer conductor. This type of cable is called a *coaxial cable*, and is illustrated in Fig. 6.5.

The two conductors can be insulated from each other either by a solid insulation along the whole length of the cable, or by insulating "spacers" fitted at regular intervals as supports for the inner conductor. The main insulation in this case is therefore the air between the two conductors.

Attenuation of Information Signals by Lines

Whatever the type of cable used, the conductors must have some electrical resistance (or opposition to current flow). Furthermore, the insulating material used to separate the two conductors of a pair will have a value of *insulation resistance*

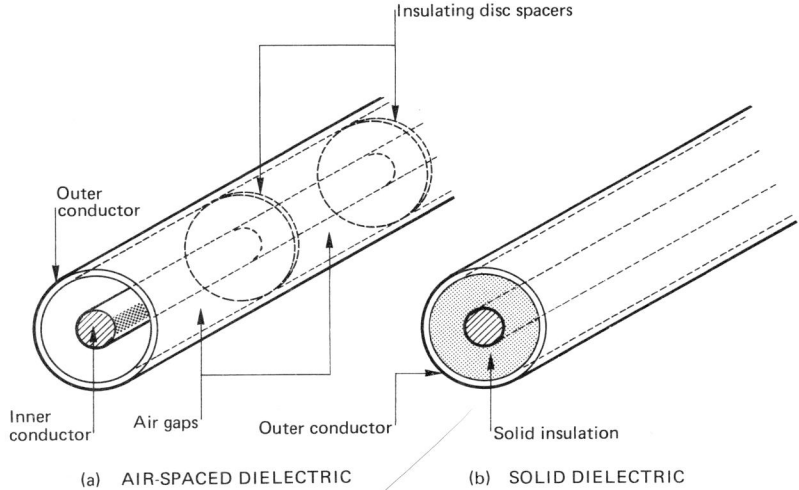

Fig. 6.5 Simple illustration of coaxial cables

(a) AIR-SPACED DIELECTRIC   (b) SOLID DIELECTRIC

which will allow a very small current to flow between the conductors instead of flowing along the conductor to the distant end.

Also, the insulation between the conductors forms a *capacitance* which provides a conducting path between the conductors for alternating currents, the conducting path becoming better as the frequency of the alternating current increases. The capacitance also has the ability to store electrical energy. This capacitive path therefore prevents part of the a.c. information signal from travelling along the conductors to the distant end of the line.

Energy is used up to make the current flow against the resistance along the conductors, and against the insulation resistance between the conductors. Energy is also used in charging and discharging the capacitance between the conductors. In multi-pair cables there is capacitive and inductive coupling between pairs, so that some energy is passed from one pair to other pairs. This reduces the amount of energy that is transmitted along the original pair, and so contributes to the loss.

In the case of an information signal, this energy is extracted from the signal source and so the energy available is gradually decreased as the signal travels along the line. This loss of energy along the line is called ATTENUATION. The *unit* used to measure this will be considered later.

If the line is long, and the attenuation is large, eventually the signal energy available at the distant end is too small to operate a receiving transducer. The attenuation generally increases as the frequency of the information signal increases, and this variation of attenuation with frequency is called *attenuation distortion.* This is illustrated simply for speech frequencies in Fig. 6.6.

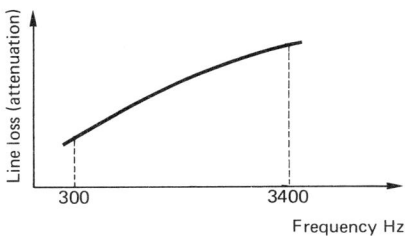

Fig. 6.6 Simple illustration of attenuation distortion of a line at speech frequencies

Noise

In any telecommunication system, whether using line or radio links, there is unwanted electrical energy present as well as that of the wanted information signal.

This unwanted electrical energy is generally called NOISE (illustrated in Chapter 1, Fig. 1.1 and Fig. 1.2) and arises from a number of different sources, which will now be considered very briefly.

(1) RESISTOR NOISE

A *conductor* is designed to carry current with minimum opposition, consistent with size and cost.

A *resistor* is a component designed to have a particular opposition to the flow of electric current in a particular circuit. This opposition is called *resistance* in d.c. circuits, but in a.c. circuits the term *impedance* is used because of added factors to be considered later. In either case the *unit* used is the OHM(Ω).

An electric current is produced by the movement of *electrons* dislodged by an externally applied voltage from the outer shells of the atoms making up the conductor material or resistor material. The movement or agitation of atoms in conductors and resistors is somewhat random, and is determined by the temperature of the conductor or resistor. The random movement of electrons brought about by thermal agitation of atoms tends to have increased energy as temperature increases.

This random movement of atoms gives rise to an unwanted electrical voltage which is called resistor noise, circuit noise, Johnson noise or thermal noise. This unwanted signal spreads over a wide range of frequencies, and the noise present in a given *bandwidth* required for a particular information signal is very important. It will also be shown later that the important noise temperature of the resistor or conductor is the Absolute or Kelvin temperature, which has its zero point at $-273°$ Centigrade. This is the temperature at which the random movement or agitation of atoms in conducting or resistive materials ceases, so unwanted noise voltages are therefore zero.

(2) SHOT NOISE

This is the name given to noise generated in active devices (energy sources), such as valves and transistors, by the random varying velocity of electron movement under the influence of externally applied potentials or voltages at appropriate terminals or electrodes.

(3) PARTITION NOISE

This occurs in multi-electrode active devices such as transistors and valves and is due to the total current being divided between the various electrodes.

(4) FLUCTUATION NOISE

This can be *natural* (electric thunderstorms, etc.) or *man-made* (car ignition systems, electrical apparatus, etc.) and again spreads over a wide range of frequencies. Such noise can be picked up by active devices and conductors forming transmission lines.

(5) STATIC

This is the name given to noise encountered in the free-space transmission paths of radio links, and is due mainly to ionospheric storms causing fluctuations of the earth's magnetic field. This form of noise is affected by the rotation of the sun (27.3 day cycle) and by the sunspot activity that prevails.

(6) COSMIC OR GALACTIC NOISE

This type of noise is also most troublesome to radio links, and is mainly due to nuclear disturbances in all the galaxies of the universe.

(7) In multi-pair cables there is capacitive and inductive coupling between different pairs which produces an unwanted noise signal on any pair because signals are transmitted to other pairs. This is called CROSSTALK between pairs and can be reduced to some extent by twisting the conductors of each pair or by changing the relative positions of pairs along the cable during manufacture or by balancing the pairs over a particular route after installation.

(8) FLICKER NOISE

The cause of this is not well understood but it is noise which predominates at low frequencies below 1 kHz, with the level decreasing as frequency increases. It is sometimes known as "excess noise" or "$1/f$ noise."

In any telecommunications system, therefore, there will be a certain level of noise power arising from all or some of the sources described, with the noise power generally being of a reasonably steady mean level, except for some noise arising from *impulsive* sources such as car ignition systems and lightning. Noise which has a sensibly constant mean level over a particular frequency bandwidth is general called *white noise*.

In order that a wanted information signal can be detected and reproduced satisfactorily at the receiving end of a system, it is essential that the power of the wanted signal is *greater* than the noise power present by at least a specified minimum value. This introduces the very important concept of *signal-to-noise ratio* in any telecommunication system as the comparison of signal power to noise power. It can be expressed simply as a *power ratio*, or more commonly it is expressed in *decibels* (dB).

The derivation of this important unit as a logarithmic ratio will be dealt with later in the course, but for now it can be simply stated that

$$\text{Signal-to-noise ratio} = 10 \log_{10} \left(\frac{\text{Signal power}}{\text{Noise power}} \right) \text{ decibels}$$

For any type of information signal there is a *minimum* acceptable value of signal-to-noise ratio for the system to operate satisfactorily. Typical *minimum* signal-to-noise ratios for different systems are as follows:

(1) Private land mobile radio telephone systems require 10 dB.
(2) Ship-to-shore radio telephone services require 20 dB.
(3) Telephone calls over the public network require 35 to 40 dB.
(4) Television systems require 50 dB.

Now, returning to the problem of sending an information signal along a line, valve or transistor amplifiers can be used to increase the signal level to compensate for the attenuation of the line. Each amplifier will generate noise internally, as previously described, so the output of each amplifier will contain the wanted signal and unwanted noise with a certain signal-to-noise ratio.

There will also be Johnson noise present on the line because of the resistance of the line conductors, and also crosstalk noise from other lines.

One amplifier *could* be placed at the sending end as shown in Fig. 6.7*a* with sufficient amplifying properties or *gain* to compensate for the line attenuation, so that the information signal reaching the other end of the line has sufficient power to operate the receiving transducer satisfactorily. This could result in a large signal power at the sending end which would cause excessive interference to other circuits in the same cable due to mutual inductance and capacitive coupling between different pairs. To avoid this problem there is a maximum permissible signal power laid down for application to pairs in different types of cable.

Another way to overcome attenuation would be to put *one* amplifier at the receiving end as shown in Fig. 6.7*b* with sufficient gain to compensate for the line attenuation. How-

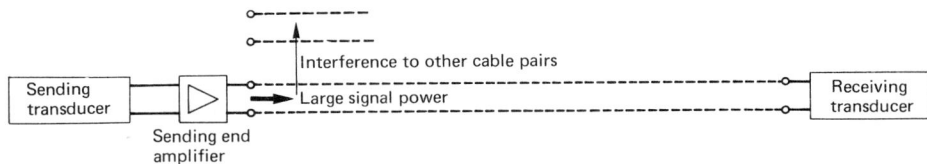

(a) SINGLE HIGH-GAIN AMPLIFIER PLACED AT SENDING END

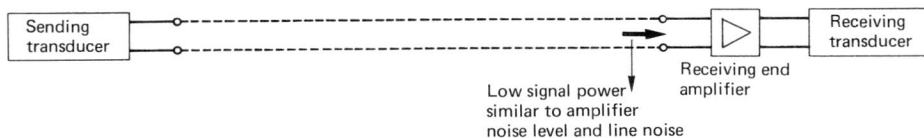

(b) SINGLE HIGH-GAIN AMPLIFIER PLACED AT RECEIVING END

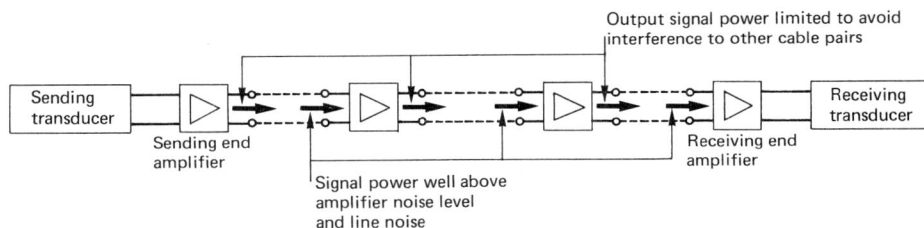

(c) LIMITED-GAIN AMPLIFIERS PLACED AT REGULAR
     INTERVALS ALONG A LINE

Fig. 6.7 Various ways of using amplifiers to overcome line attenuation in a one-way telecommunication system

ever, if the line is long with a restricted permissible power at the sending end, the attenuation could be such that the information signal power at the receiver is low enough to give an inadequate signal-to-noise ratio when the line noise and noise generated by the receiver are considered.

To overcome these problems, amplifiers must be placed at regular points along the line where the information signal power is still large enough to give an adequate signal-to-noise ratio compared with the amplifier noise and line noise. This simple concept is illustrated in Fig. 6.7c.

Since an amplifier is generally a one-way device with definite input and output connections, the arrangement illustrated in Fig. 6.7 needs to be duplicated to enable information signals to be transmitted in the opposite direction. However it has previously been seen that simple telephone communication circuits carry information in both directions over a single pair of wires. To meet this requirement, it is therefore necessary to

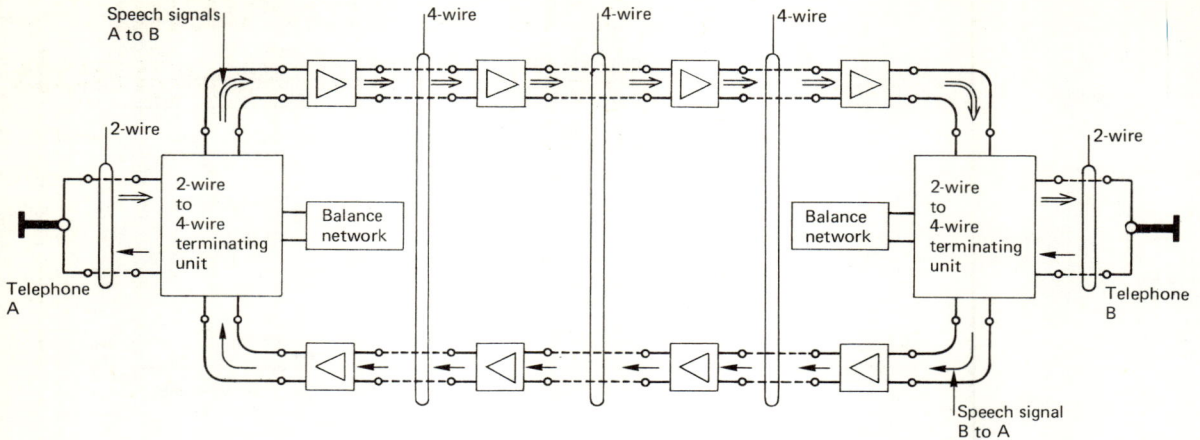

Fig. 6.8 Illustration of 4-wire amplified telephone circuit

arrange that when amplification is needed over telephone circuits, the normal simple two-wire connection is changed into a 4-wire connection to provide one pair for transmitting signals in each direction. This arrangement will be dealt with in more detail later in the course, but the simple principles are illustrated in Fig. 6.8.

It should be added here that there are certain types of amplifier that can be inserted into a 2-wire line to give amplification in both directions, but the use of these in the public telephone network is limited.

7 Introduction to Public Telephone and Telegraph Networks

Introduction

Fig. 5.2 (p. 38) illustrated very simply the principle of a telephone connection, with the telephone instrument at each end of the link having sending and receiving transducers known as the telephone transmitter and telephone receiver respectively. This can be represented even more simply as shown in Fig. 7.1.

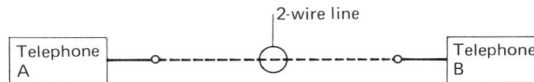

Fig. 7.1 Simple telephone system

Another form of communication system, which was mentioned in Chapter 1, deals with the written word instead of the spoken word, and is called a telegraph system. One type of telegraph system has two teleprinters connected together by a line.

A TELEPRINTER is a device that looks like a large typewriter, having a keyboard with letters, figures and other commonly-used characters. When any key is depressed, the teleprinter mechanism produces an electrical signal that represents the particular character in the form of a series of voltage pulses at +80 V and −80 V in accordance with a 5-unit or Murray code. The voltage pulses are applied to the line that is connected to another teleprinter so that a series of current pulses flows along the line to the other teleprinter. The current pulses energise an electromagnetic receiver which then operates the teleprinter mechanism to print the character that was originally selected at the sending teleprinter.

At the same time, at the sending teleprinter the signal pulses can be connected internally to the receiving electromagnet of the teleprinter, so that a copy is made of the message being sent to the distant teleprinter. The message can be produced on a normal sheet of paper (page-printing) or on a continuous paper tape. The latter method is used in the public telegraph network so that the tape can be cut up and stuck on to the telegram forms which are eventually delivered to the destination.

Fig. 7.2 illustrates a simple teleprinter connection. The line connecting the two teleprinters can be either a single wire or a 2-wire line, depending on the particular system being considered.

Fig. 7.2 Simple telegraph system

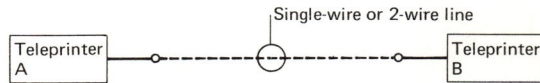

The two communication systems illustrated in Figs. 7.1 and 7.2 show permanent connections between two points A and B, and represent the very early uses of these two methods of communication. It is not difficult to imagine that, in the very early days of telephone and telegraph communication, other people would want to be included in such systems, and that each person or *subscriber* to a system would wish to have access to all other subscribers when necessary.

Fig. 7.3 Fully interconnected communication system

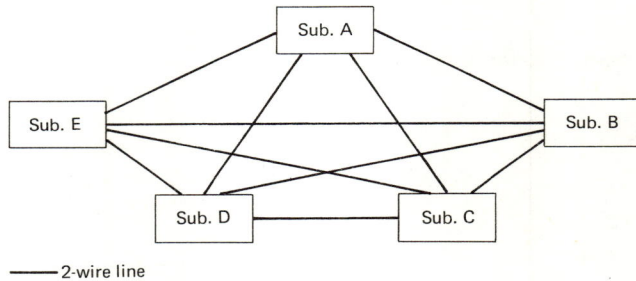

If we consider a small number, say four or five, wishing to set up a telephone or telegraph communication system, it is possible to arrange complete interconnection by providing suitable line connections between all the telephone or telegraph subscribers as shown in Fig. 7.3. In such a small system it is necessary to provide some form of signalling code so that each subscriber can call any other subscriber when desired.

It should also be fairly obvious that there is a limit to the size of such a fully-interconnected system, both in terms of number of subscribers and in the geographical area that can be covered. One can imagine the problems of connecting one subscriber to hundreds or thousands of other subscribers over long distances.

It was a logical development therefore to provide a central point to which all subscribers are connected, and where any two subscribers are interconnected on demand. This central point is called an EXCHANGE, because connections can be exchanged when required, and we therefore see the introduction of telephone and telegraph exchanges as switching centres, with operators employed to do the switching on demand by calling subscribers.

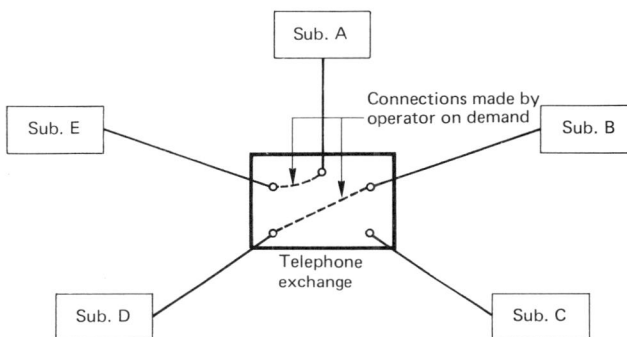

Fig. 7.4 Simple telephone system with a central switching point or exchange

The principle of a telephone exchange system is illustrated in Fig. 7.4.

Once a telephone or telegraph system had been set up in a particular country, many other subscribers would wish to be included, and sooner or later the problem of providing a national network would arise. Considering Fig. 7.4 again, it should be clear that it would be impracticable to provide *one* central telephone or telegraph exchange for a country with all subscribers connected to it, because of the amount of switching equipment that would be necessary, and because of the length of line needed for each subscriber. So, a large number of exchanges are needed in a NATIONAL NETWORK, with each exchange having clearly defined geographical boundaries and a limited number of subscribers connected to it.

The problem then arises as to how to enable subscribers on one telephone exchange to be connected to any other subscriber in the country. Just as it is impracticable to fully-interconnect subscribers on one exchange as shown in Fig. 7.3, it is clearly just as impracticable to fully-interconnect all exchanges. This is demonstrated in Fig. 7.5, with only five fully-interconnected exchanges.

The lines interconnecting the exchanges represent a sufficient number of 2-wire links to handle the calls between subscribers from any two exchanges. These interconnecting lines are called JUNCTIONS. Clearly if this idea was extended to a large number of exchanges, the number of lines needed would be immense. This problem can be tackled by arranging telephone exchanges in a particular geographical area into a

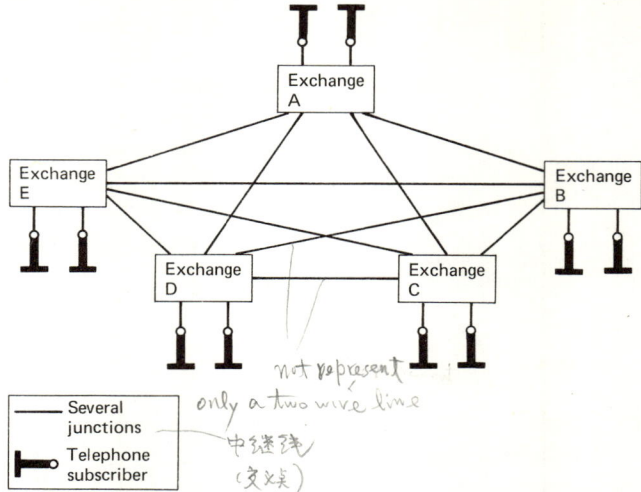

Fig. 7.5 Five telephone exchanges in a fully interconnected group

Annotations: not represent only a two wire line / 中继线 (交换)

GROUP of exchanges, and selecting *one* of the exchanges as a switching centre for the group and called a GROUP SWITCHING CENTRE (GSC). This is illustrated in Fig. 7.6.

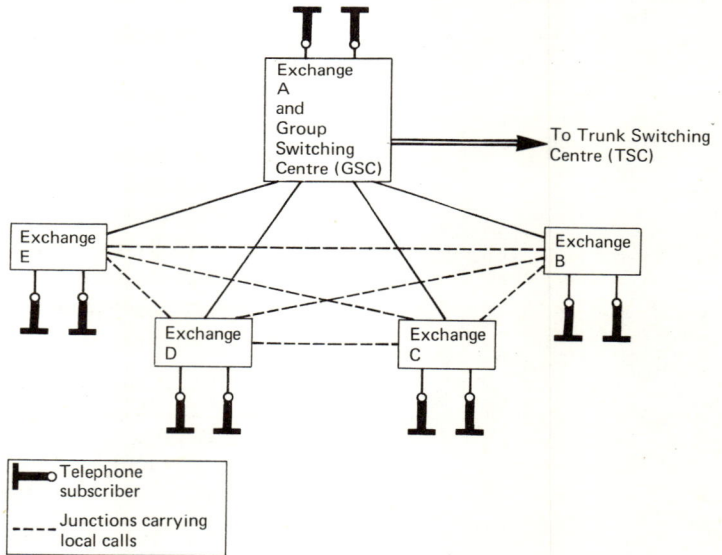

Fig. 7.6 Concentration of trunk calls at a group switching centre

Connections can be made within a group of local exchanges by a fully-interconnected junction network, but all connections to subscribers *outside* the group are routed via the GSC. In this way a number of local exchanges share one GSC, and calls outside the group are therefore concentrated at the GSC.

In the same way, a number of groups can share one switching centre, for routing long-distance calls, as illustrated in Fig. 7.7. The switching centres for long-distance trunk calls are called TRUNK SWITCHING CENTRES (TSC). Also shown in Fig. 7.7 are *direct* routes between adjacent GSCs where the number of calls justifies such provision.

认为...合理

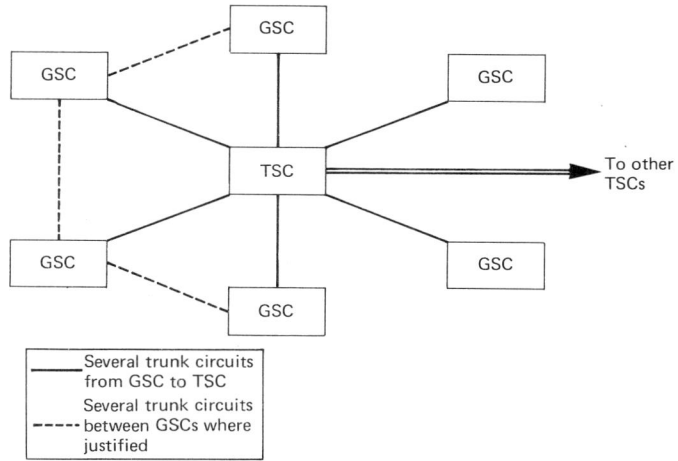

Fig. 7.7 Concentration of trunk calls at a trunk switching centre

In this way, by setting up a graded network or hierarchy of switching centres, the whole of a country can be included with a minimum number of interconnecting junctions and trunk circuits, with calls being collected or concentrated at strategic points.

When the national network in the UK was re-planned in the 1930s, switching at the various centres was done by the telephone operators working manual switchboards. It is now possible to dial directly to nearly all distant subscribers by the Subscriber Trunk Dialling (STD) network, the provision of which is now effectively complete. The STD network is divided into two sections:

(1) One section handles calls over relatively short distances, and also long-distance calls that require only one intermediate switching centre between originating and objective GSCs. The switching of circuits in this part of the network is done in the two-wire portion of the circuits.

(2) The other section, called the TRANSIT network, is used for long-distance calls that require more than one intermediate switching centre. The switching is done in the 4-wire portion of the circuits, and a high-speed signalling system is used.

The principle of the national telephone switching network in the UK is illustrated in Fig. 7.8. The TSCs are divided into two categories, DISTRICT SWITCHING CENTRES (DSC) and MAIN SWITCHING CENTRES (MSC), with the MSCs being fully interconnected. Each GSC is connected to at least *one* DSC.

So, in the national network there are various grades of exchange in order of importance. The more important a particular grade of exchange is, the fewer there are. There are also various grades of interconnecting line, for example subscribers lines, junctions and trunks, and again the number provided gets fewer as length, importance and cost increase.

An illustration of how two telephone subscribers in different parts of the country could be connected together is given in Fig. 7.9.

For calls between two subscribers relatively close to each other, the transit network is not needed, as shown in Fig. 7.10. The intermediate GSC may or may not be needed, according to the distance between subscribers.

The next requirement for a subscriber may be to make a telephone call to another country. In the same way as national calls are progressively concentrated at GSCs, DSCs and MSCs, so calls to other countries are concentrated at an INTERNATIONAL EXCHANGE in any country. In the UK the international exchange is located in London. Fig. 7.11 illustrates a connection to the international exchange.

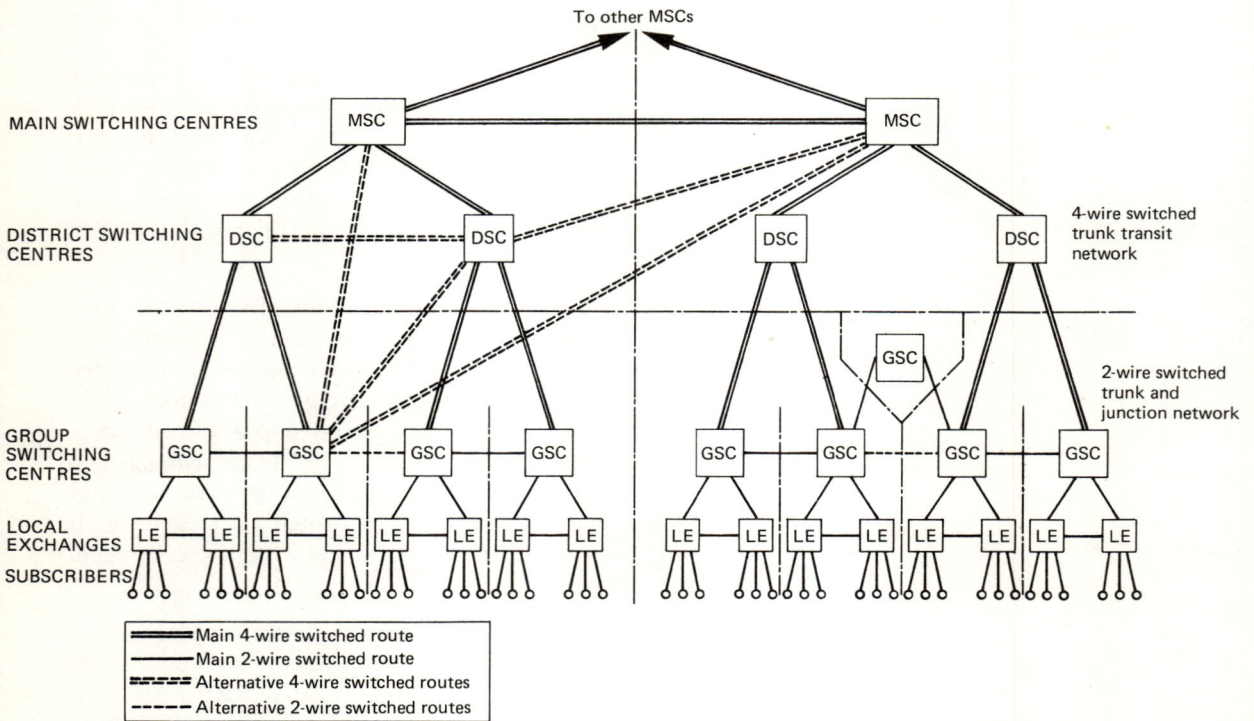

Fig. 7.8 Outline of a national telephone switching network

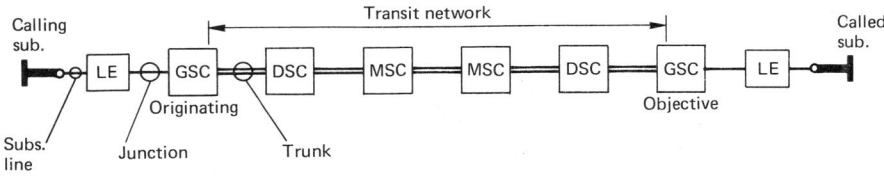

Fig. 7.9 Typical connection between two subscribers via transit network

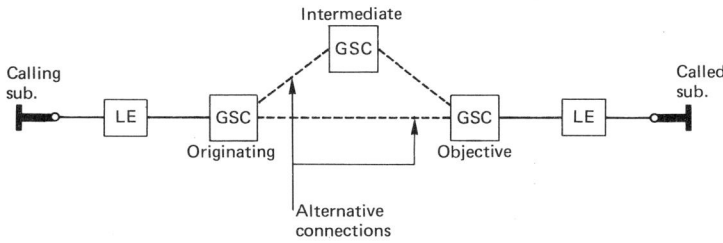

Fig. 7.10 Connection between two subscribers without using transit network

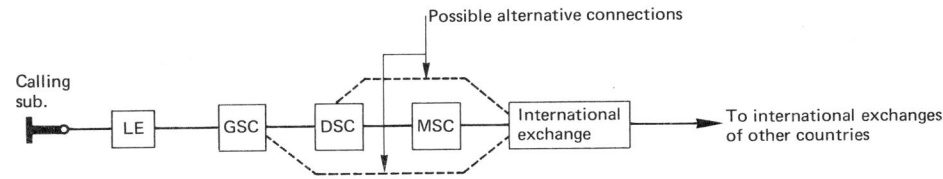

Fig. 7.11 Typical possible connections from subscribers to international exchange

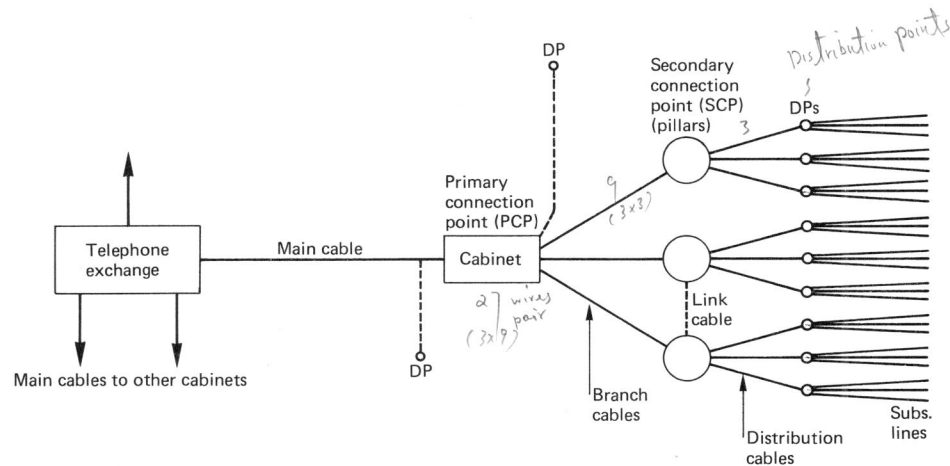

Fig. 7.12 Principle of local telephone line distribution network

Local Distribution Network

The way in which subscribers' telephone instruments are connected to the local telephone exchange will now be considered, bearing in mind that each telephone requires a two-wire line or PAIR.

These pairs leave the exchange in large multi-pair *main* cables which feed primary connection points or cast-iron *cabinets* placed at various points in the telephone exchange area.

Each PCP or cabinet is then connected to a number of secondary connection points or *pillars* by means of smaller multi-pair *branch* cables. From the SCPs or pillars, distribution cables are connected to *distribution points* (DPs), each of which will feed a small number of subscribers' premises.

The connection to each subscriber from the DP may be a simple drop-off insulated pair, or *may* include a span or two of overhead bare wires, although this method has virtually disappeared as local distribution networks are brought up to date. The cabinets and pillars give flexibility in distribution of cable pairs throughout the exchange area.

This system of local line distribution has been a standard installation for a number of years, but is being replaced by a similar flexible system using cast-iron cabinets instead of the original concrete-type pillars for the SCPs, with a different method of connecting pairs together inside the cabinets. This is illustrated in Fig. 7.12.

Connections Between Telephone Exchanges

In Fig. 7.8 the national network is illustrated, with local subscribers' lines, junctions between certain exchanges, and trunk circuits between other exchanges. In Chapter 6 the different types of cable were mentioned briefly.

Generally, for the low-category lines from subscribers to local exchanges, and for junctions between local exchanges and GSCs, MULTI-PAIR types of cable are used. Between GSCs and TSCs, and between TSCs, where many originating calls have been concentrated, COAXIAL-type cables will generally be used.

These cables are generally buried in the ground, but occasionally submarine cables may be required where the cable has to cross rivers, lakes or estuaries, for example.

Supervisory Signals

When national telephone networks were first introduced, the switching between lines was done by telephone operators working at manual switchboards. It was soon realized that there are several advantages in providing automatic equipment which is remotely controlled by calling subscribers using a dial incorporated in the telephone instrument.

In the manual system, when a subscriber wishes to make a call, a signal is sent to call the operator. The calling subscriber knows when the operator has answered because the operator says "number please". In the automatic system, some distinctive form of signal must be passed to the calling subscriber to state that the automatic equipment is ready to receive routing information from the dial. This is called DIAL TONE.

In the manual system, when the operator *says* "number please," the calling sub gives the number required. The operator tests the wanted line, and if it is free, she sends a ringing current to ring the bell of the wanted telephone, and she *tells* the calling sub that she is trying to connect the call. To convey the same information in an automatic system a recognisable signal is needed, and this is called RING TONE. At the same time RINGING CURRENT is applied to the called subscribers line to ring the telephone bell, just as the operator does in the manual system.

In the manual system, if the wanted subscriber is already engaged on a call, the operator *tells* the calling subscriber that the line is engaged. In an automatic system, a BUSY TONE is needed to give the same information. If a calling subscriber asks the operator in a manual system for a number that is out of order, or is not a working line, then she *tells* the calling subscriber that this is so. In an automatic system a NUMBER UNOBTAINABLE (NU) TONE is used to convey this information to the calling subscriber.

In an automatic exchange, when a calling subscriber dials a wanted number, the automatic equipment *may* not be able to connect the call because certain sections may all be engaged on other calls. This information must be conveyed to the calling subscriber by means of an EQUIPMENT BUSY TONE.

Telegraph Networks

In most countries a need has emerged for two teleprinter networks.

One is a *public* network that enables people to send *telegrams* to any address in the country. Such telegrams are originated either by handing in a *written* telegram form at a Post Office, or by *telephoning* the telegram to a special telegraph switchboard operator by dialling a particular code. The telegram message is then conveyed by teleprinter to the telegraph terminal nearest the destination. The telegraph operator then sticks the teleprinter tape message on to a telegram form, which is then delivered to the destination address.

The second need for a teleprinter network arises from private communication between individuals or firms and organizations where a *printed record* of the communication is preferred to a *spoken* message by telephone. Such a system is known as TELEX.

In the early days of private teleprinter communication, the public telephone network was used. Each subscriber who needed the facility rented a teleprinter as well as a telephone. To send a Telex message, a telephone call was made to the wanted number, and when the connection was established both subscribers switched from telephone to teleprinter, and the message was transmitted. This system is illustrated in Fig. 7.13.

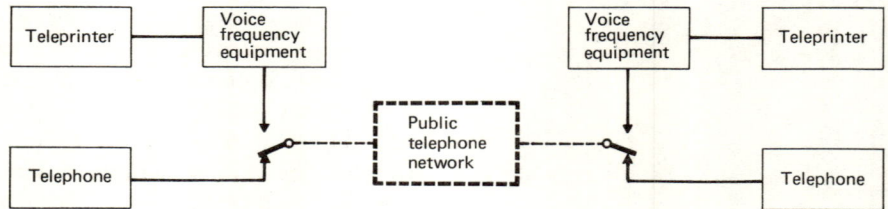

Fig. 7.13 Original arrangement of telex network

The voice-frequency equipment shown in Fig. 7.13 is needed to change the ±80 V coded d.c. pulses into alternating voltage having a frequency within the commercial speech range, so that the telephone line can handle the teleprinter information signals. Several problems arose from this arrangement, and it was decided to set up a completely separate switching network for the Telex Service, which was completed in the UK in 1954.

Fig. 7.14 Automatically switched telex network

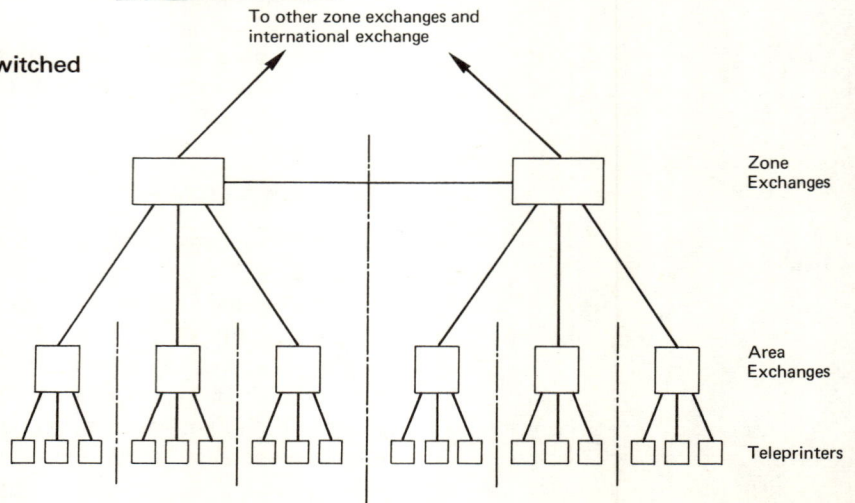

This separate Telex network originally used exchanges with manual switching by Telex operators, but these exchanges were eventually replaced by approximately 50 automatic Telex exchanges. Each Telex exchange serves a certain area, and a number of areas form a Zone. There are six zones, which are fully interconnected. The network is illustrated in Fig. 7.14.

8 Introduction to Telephone Exchange Switching Principles

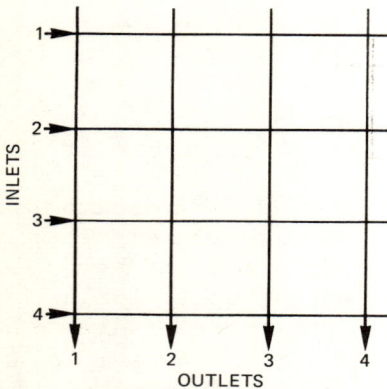

Fig. 8.1 Simple 4×4 matrix switch (see Fig. 8.8 for typical crosspoint connections)

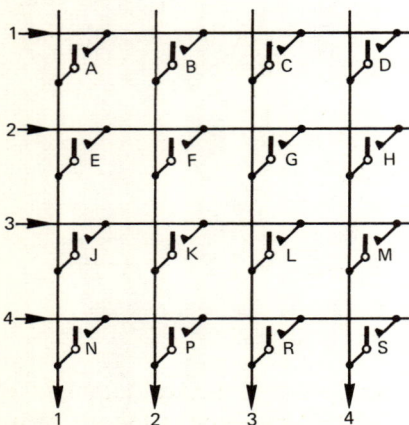

Fig. 8.2 Simple principles of switching by a 4×4 matrix switch

In Chapter 7 the need for telephone and telegraph exchanges in a national switching network was introduced. We now have to consider some different ways in which the switching between lines is achieved in exchanges.

Matrix Switching

One method is to use a MATRIX SWITCH, the principle of which can be explained by considering the circuits which are to be connected together as being arranged at right angles to each other in horizontal and vertical lines. These lines represent inlets and outlets of the switch. This idea is illustrated in Fig. 8.1.

The intersections between horizontal and vertical lines are called CROSSPOINTS. At each crosspoint some form of switch contact is needed to complete the connection between horizontal and vertical line, as shown in Fig. 8.2.

Any of the 4 inlets can be connected to any of the 4 outlets by closing the appropriate switch contacts. For example,

(a) Inlet 1 can be connected to outlet 2 by closing contact B.
(b) Inlet 4 can be connected to outlet 3 by closing contact R.

Considering Figs. 8.1 and 8.2 again, it can be seen that with 4 inlets and 4 outlets there are 16 crosspoints. Obviously, the number of crosspoints in any matrix switch can be calculated by multiplying the number of inlets by the number of outlets. This is further illustrated in Fig. 8.3.

If there are n inlets and m outlets, then the number of crosspoints is $(n \times m)$.

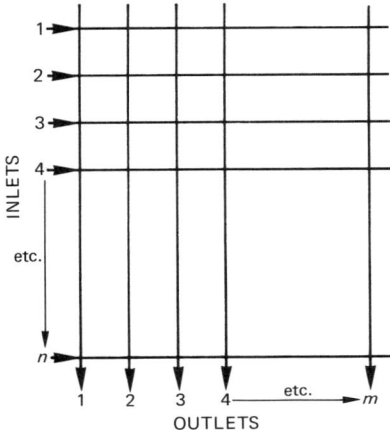

Fig. 8.3 Number of crosspoints in a matrix switch

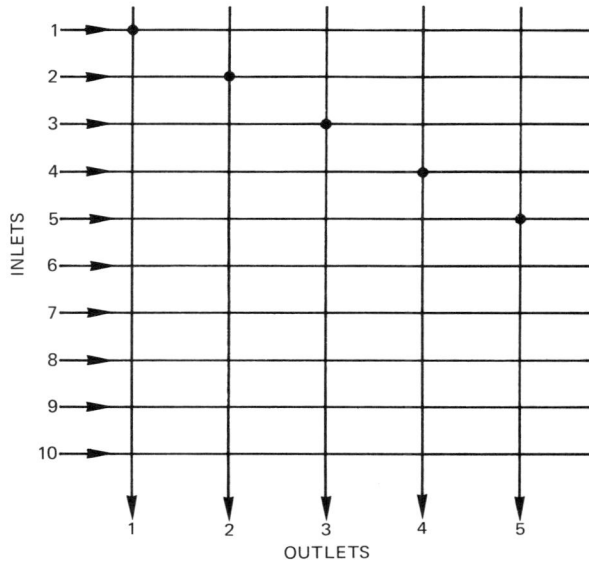

Fig. 8.4 Simple 10×5 matrix switch

(1) If n is larger than m, that is if there are more inlets than outlets, then not all the inlets can be connected to a different outlet. When all the outlets have been taken there will be some inlets still not in use.

(2) If m is larger than n, that is there are more outlets than inlets, then when all inlets are each connected to an outlet, there will be some outlets still not in use.

So, the maximum number of simultaneous connections that can be carried by a matrix switch is given by whichever of the number of inlets or outlets is smaller. For example, if there are 10 inlets and 5 outlets, then the maximum number of simultaneous connections possible is 5, as illustrated in Fig. 8.4.

In Fig. 8.4,

inlet 1 is connected to outlet 1
inlet 2 is connected to outlet 2
inlet 3 is connected to outlet 3
inlet 4 is connected to outlet 4
inlet 5 is connected to outlet 5

Efficiency

The suitability of a matrix switch as previously described is sometimes measured in terms of the efficiency in the use of its crosspoints. Take a simple matrix switch with 4 inlets and 4 outlets as illustrated in Fig. 8.5. There are 16 crosspoints, but

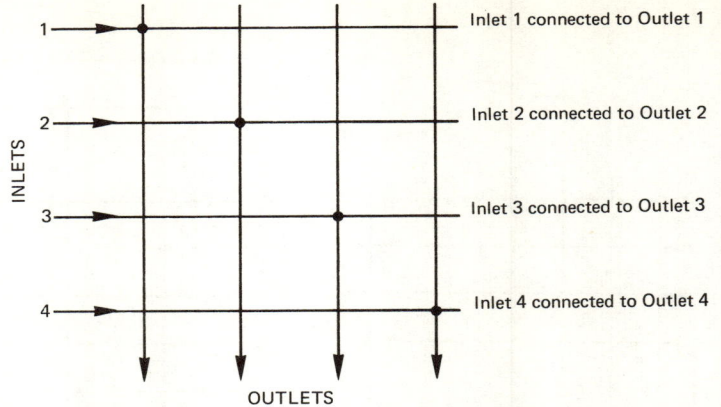

Fig. 8.5 Maximum number of simultaneous connections in a 4 × 4 matrix switch

only 4 can be in use at any one time, when the 4 inlets are connected to the 4 outlets.

The efficiency in use of crosspoints of the matrix switch is calculated by

$$\frac{\text{Maximum number of crosspoints in use simultaneously}}{\text{Total number of crosspoints in the matrix}} \times 100\%$$

In the case shown in Fig. 8.5,

$$\text{Efficiency} = \frac{4}{16} \times 100\% = \frac{100}{4}\% = 25\%$$

The efficiency of this type of matrix switch gets smaller as the switch gets larger. For example, consider a matrix switch with 15 inlets and 15 outlets, as shown in Fig. 8.6.

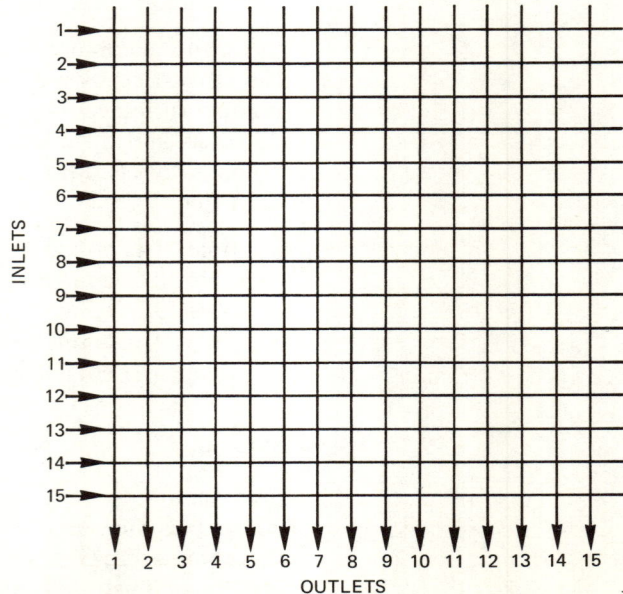

Fig. 8.6 15 × 15 matrix switch with 225 crosspoints

The total number of crosspoints $= 15 \times 15 = 225$.

Maximum number of crosspoints in use simultaneously is 15.

$$\text{Efficiency} = \frac{15}{225} \times 100\% = \frac{15}{2.25}\% = 6.67\%$$

This low efficiency can be improved for the same number of inlets and outlets by arranging the switch in *two* stages instead of one, using small basic matrix switches. This is illustrated in Fig. 8.7, where six 5×3 basic matrix switches are arranged to give a two-stage 15×15 switch. The two stages are connected together by 9 links.

In Fig. 8.7, total number of crosspoints is $6 \times 5 \times 3 = 90$.

Since there are only 9 links between the A and B switches, only 9 inlets can be connected to 9 outlets at any one time.

Each connection from an inlet to an outlet uses *two* crosspoints, one in an A-switch and one in a B-switch.

So, maximum number of crosspoints in use simultaneously is 18. Therefore

$$\text{Efficiency} = \frac{18}{90} \times 100\% = \frac{180}{9}\% = 20\%$$

This is a much greater efficiency than is given by the single-stage 15×15 matrix shown in Fig. 8.6.

However, there is also a disadvantage in using this arrangement. Clearly, from Fig. 8.7, the 15 inlets are divided into three groups of 5 at the A-switches, and only 3 of each group of 5 inlets can be connected to a B-switch at any one time, where they are then connected to any 3 of the 15 possible outlets.

Fig. 8.7 Two-stage 15×15 matrix switch using link trunking

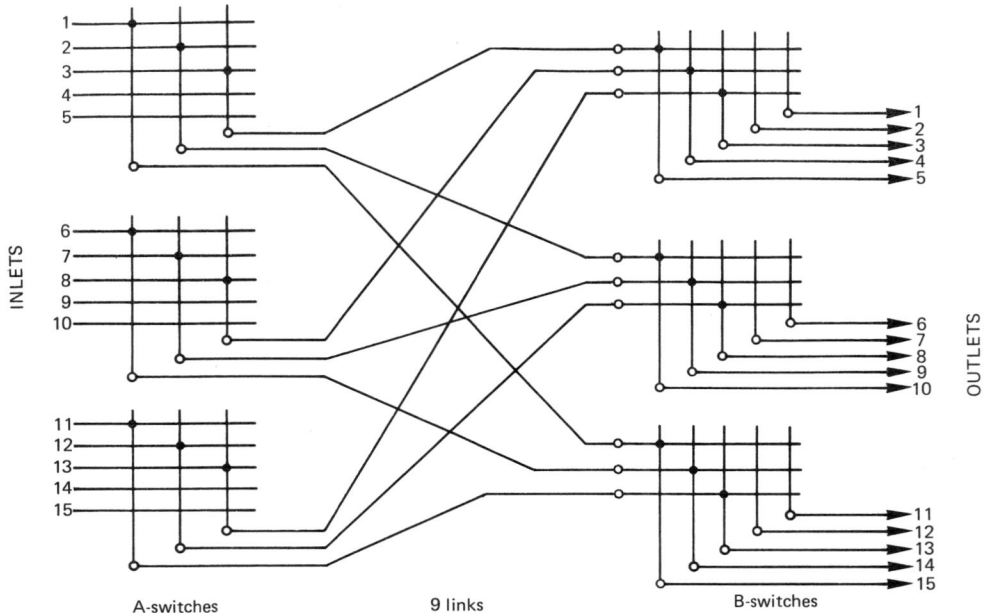

Fig. 8.7 shows the following interconnections:

inlet 1 to outlet 15
inlet 2 to outlet 10
inlet 3 to outlet 5
inlet 6 to outlet 14
inlet 7 to outlet 9
inlet 8 to outlet 4
inlet 11 to outlet 13
inlet 12 to outlet 8
inlet 13 to outlet 3

So, inlets 4, 5, 9, 10, 14 and 15 cannot be connected to any outlet, even though outlets 1, 2, 6, 7, 11 and 12 are free, because all 9 links are already in use.

Further, even with only *one* connection made, say between inlet 1 and outlet 15 as shown, then inlets 2, 3, 4 and 5 cannot be connected to any of the outlets 11, 12, 13 or 14 because the *one link* between the particular A and B switches is already in use.

This is called INTERNAL BLOCKING, or LINK CONGESTION and must be considered when designing multi-stage matrix switches.

In telephone exchanges it is necessary to connect the two wires from one telephone to the two wires of another telephone. It is also necessary to be able to *guard* the calling line and the called line so that neither can be seized by another subscriber. Generally a third wire called the PRIVATE WIRE and designated P-wire is used for this purpose. There may also be need for a fourth wire or connection for switching and holding purposes. So, each individual *crosspoint* in a matrix may consist of 4 input wires to be connected to 4 output wires.

This is illustrated in Fig. 8.8 which shows a typical crosspoint arrangement in a TXE.2 electronic-controlled exchange that uses reed relays as the crosspoint switches. (A brief description of a reed relay is given at the end of this chapter.)

To connect inlet 1 to outlet 1, a current is passed through the reed relay coil (RL), and the 4 contacts operated by the reed relay connect the $-$, $+$, P and H wires of the inlet to the $-$, $+$, P and H wires of the outlet respectively.

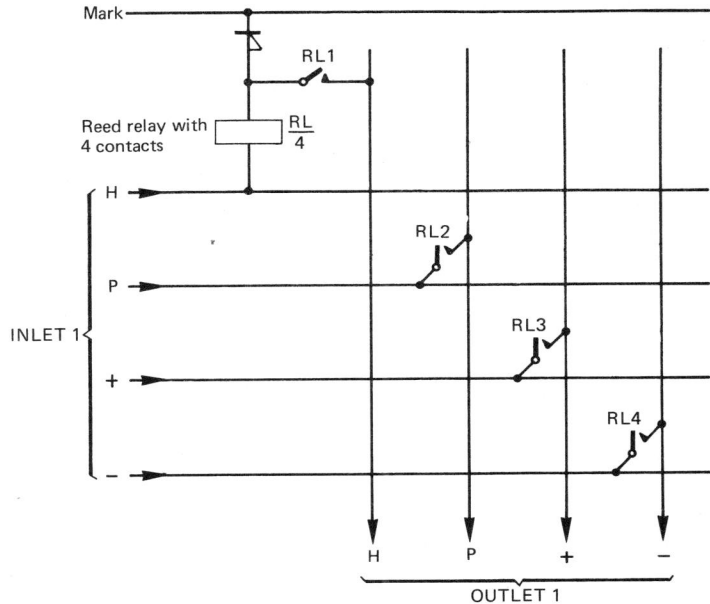

Fig. 8.8 Typical application of matrix switch crosspoint connection

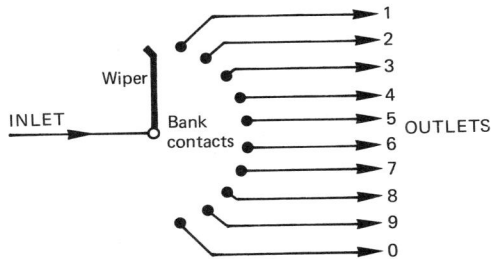

Fig. 8.9 Simple principle of switching by electro-mechanical uniselector of one-from-ten outlets

Step-by-Step Switching

This is another method of exchange switching used for many years in several countries, including the UK. The selection of a particular line is based on a one-from-ten selection process. For example, Fig. 8.9 shows a simple switch that has ten contacts arranged around a semi-circular arc or BANK, with a rotating contact arm or WIPER that can be made to connect the inlet to any one of the ten bank contact outlets as required. The wiper is rotated by a simple electro-magnet driving a suitable mechanism, so the arrangement is called an ELECTRO-MECHANICAL SWITCH.

The wiper rotates in one direction only, so this type of electro-mechanical switch is called a UNISELECTOR. Clearly the inlet can be connected to any one of the ten outlets, but the outlets are numbered from 1 to 0, which is normal practice in the step-by step switching system.

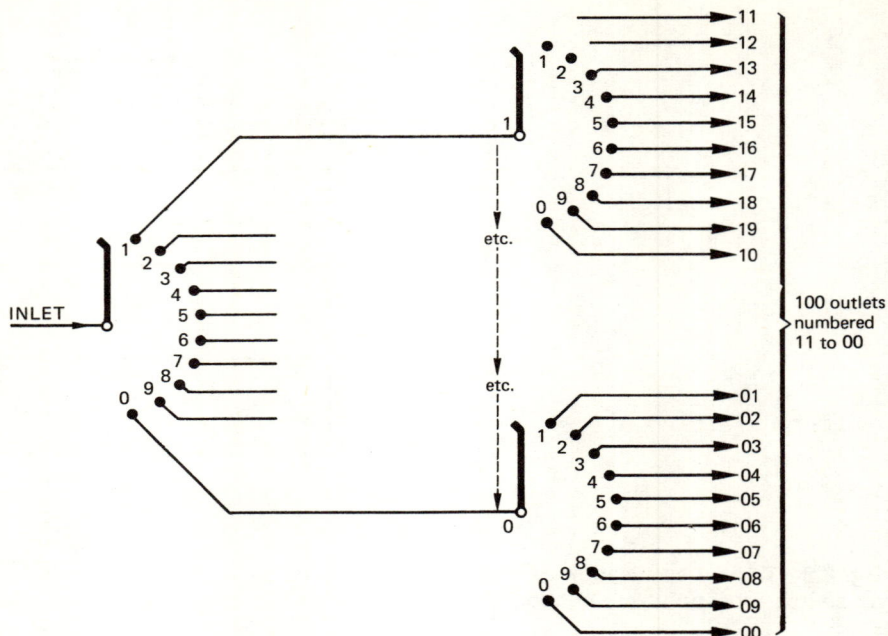

Fig. 8.10 Simple step-by-step selection of one-from-a-hundred outlets

This principle can be extended to enable the inlet to be connected to any one from 100 outlets by connecting each of the ten outlets of the first uniselector to the inlet of another uniselector, as shown in Fig. 8.10. The switching of the inlet to any one of the 100 outlets (numbered 11 to 00) is done in *two steps*, the first digit being selected on the first uniselector, and the second digit being selected on the second uniselector.

If each of these 100 outlets is now connected to another uniselector, the inlet can then be connected to any one from 1000 outlets, numbered from 111 to 000, with the digits being selected one at a time on the three successive switching stages. This arrangement can theoretically be extended to accommodate any number of digits in a particular numbering scheme.

The same sort of numbering scheme can be provided (on a step-by-step basis) by a different type of electro-mechanical switch called a TWO-MOTION SELECTOR. The principle is illustrated simply in Fig. 8.11.

The bank of fixed contacts now contains 10 semi-circular arcs, each having 10 contacts, and arranged above each other. The moving contact or wiper can be connected to any one of the 100 bank contacts by first moving *vertically* to the appropriate level, and then rotating *horizontally* to a particular con-

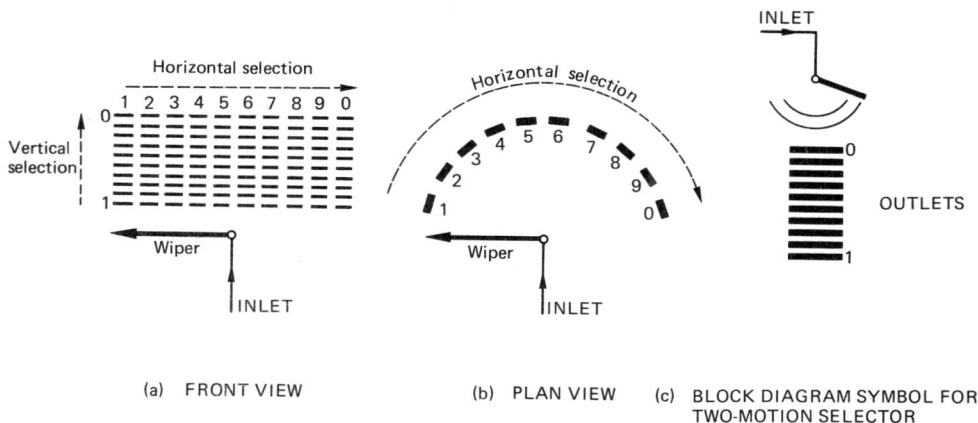

(a) FRONT VIEW   (b) PLAN VIEW   (c) BLOCK DIAGRAM SYMBOL FOR
                                         TWO-MOTION SELECTOR

Fig. 8.11 Simple principle of one-from-a-hundred selection by two-motion selector

tact on that level. The 100 outlets are numbered from 11 to 00.

The diagram symbol used to illustrate the 100-outlet 2-motion selector is shown in Fig. 8.11c.

As with the uniselector arrangement, the two-motion selector system can be extended to give access to any number of outlets by adding an extra switching stage for each extra digit required in the numbering scheme. A 3-digit numbering scheme from 111 to 000 outlets is illustrated in Fig. 8.12, with the one-from-a-hundred selector preceded by a one-from-ten selector.

In Fig. 8.12 the first digit of the 3-digit numbering scheme raises the *wiper* of the first 2-motion selector to the appropriate *vertical* level. The selector then automatically searches for a *free* outlet *on that level* to the next selector which caters

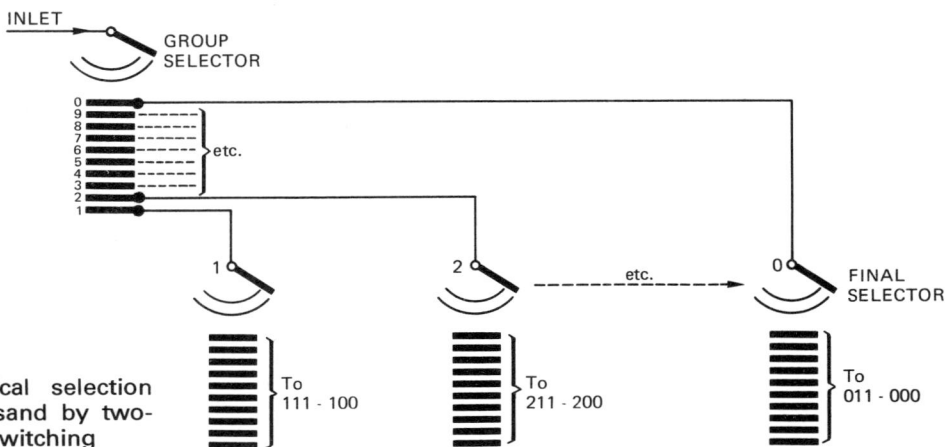

Fig. 8.12 Theoretical selection of one-from-a-thousand by two-stage step-by-step switching

for the last *two* digits of the 3-digit numbering scheme, as illustrated in Fig. 8.11.

The arrangement of the outlets from *one* level of the first stage of selection is illustrated in Fig. 8.13. The second stage of switches connected from level 5 of the first stage *shares* the 100 possible outlets 511 to 500. The method of sharing or paralleling the outlets from a number of selectors in a *multiple* is illustrated in Fig. 8.14.

The two-motion selectors of the first stage, which select a particular *level* according to the *first* digit of a 3-digit numbering scheme, are called GROUP SELECTORS.

The two-motion selectors of the *second* stage which handle the last *two* digits of a 3-digit numbering scheme are called FINAL SELECTORS.

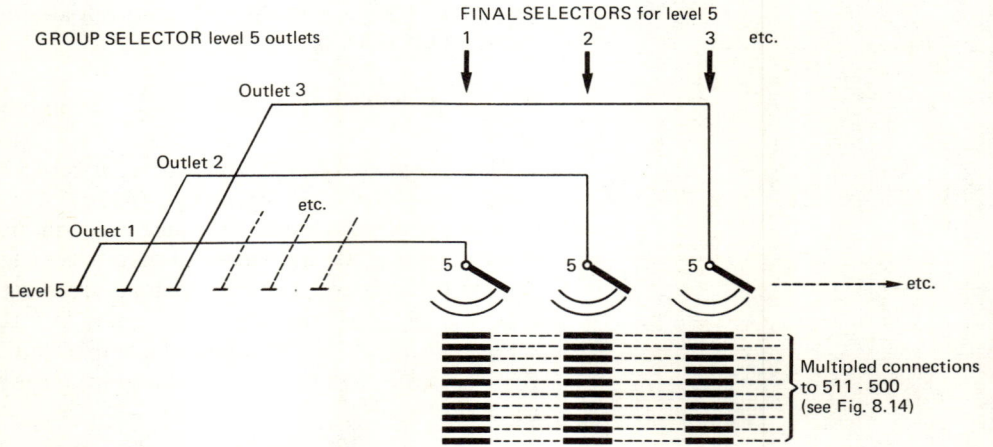

Fig. 8.13 Outlets to 511–500 paralleled or multipled from several selectors

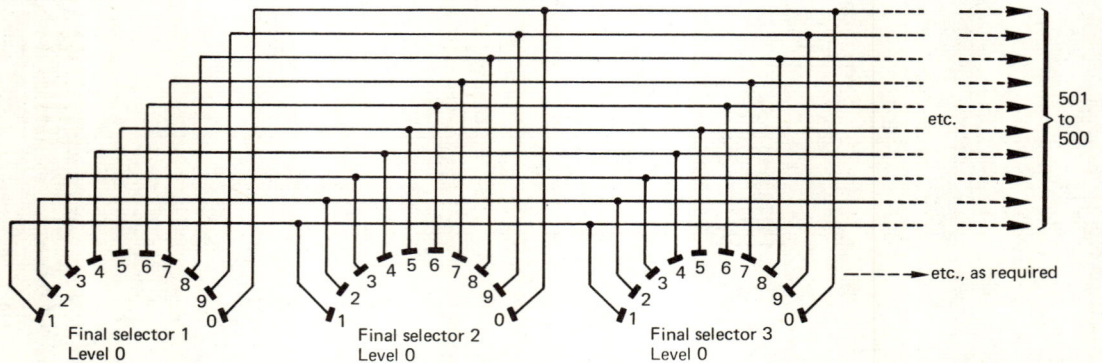

Fig. 8.14 Illustration of multipled outlets from level O of Final selectors in Fig. 8.13
(Final selectors connected from level 5 of group selectors as in Fig. 8.13)

In order to provide access to 10 000 lines a further stage of group selectors is added *before* the final selectors. Fig. 8.15 illustrates how a calling subscriber can be connected to other subscribers in an exchange having a 4-digit numbering scheme.

Theoretically, a 4-digit numbering scheme can accommodate 10 000 subscribers, but it is necessary also to provide junctions to other exchanges, junctions to the GSC for STD calls, lines to the operator and other Enquiry Services, lines to telegrams, and so on. This means that only levels 2, 3, 4, 5, 6 and 7 are usually available on the first group selectors for connection to other subscribers via second group selectors and final selectors. So the capacity of the exchange is reduced to 6000 subscribers instead of the theoretical number of 10 000.

The usual arrangement for the first group selector levels is shown in Fig. 8.16.

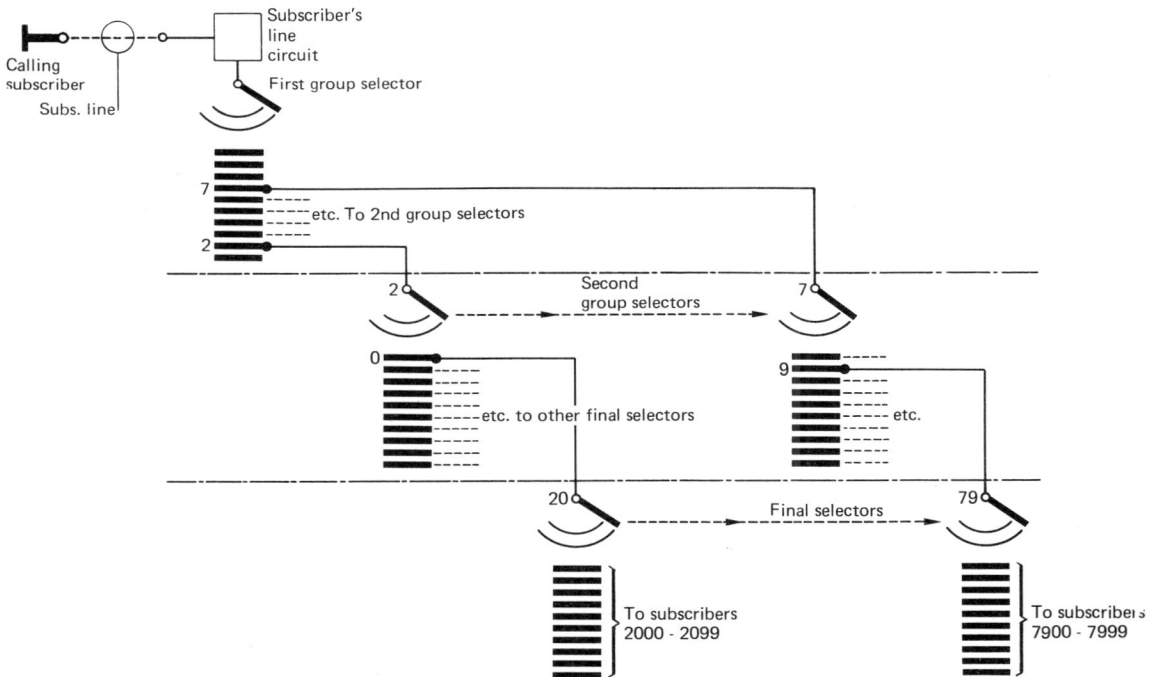

Fig. 8.15 Simple trunking diagram of 4-digit step-by-step automatic exchange

Fig. 8.16 Typical facilities available from first group selector levels

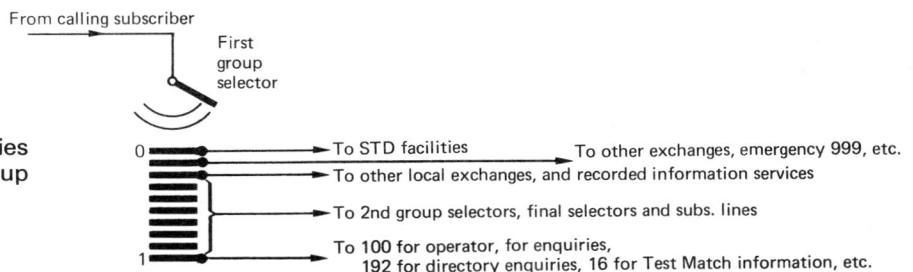

The Reed Relay

A relay is a device for remotely operating switches to control the flow of current in other electrical circuits. A reed relay is one particular type of relay used amongst other applications in the control circuits of electronic telephone exchanges. It is based on the fact that an electric current passing through a coil of wire produces an electromagnet, with the ends of the coil having opposite magnetic polarities, as illustrated in Fig. 8.17.

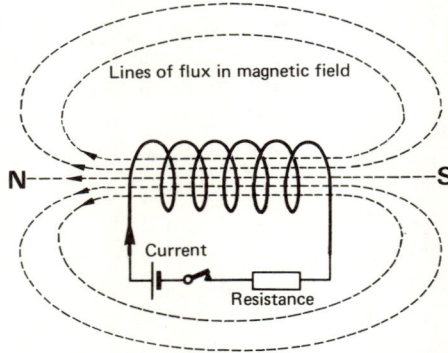

Fig. 8.17 Coil of wire as a simple electromagnet

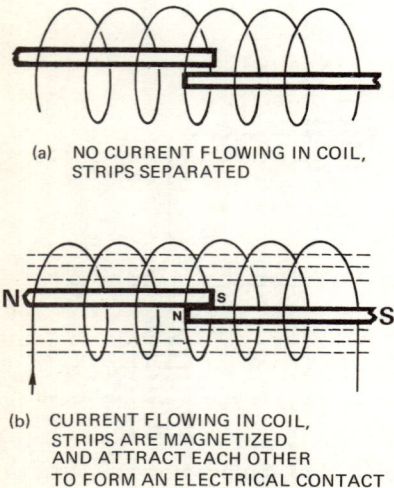

(a) NO CURRENT FLOWING IN COIL, STRIPS SEPARATED

(b) CURRENT FLOWING IN COIL, STRIPS ARE MAGNETIZED AND ATTRACT EACH OTHER TO FORM AN ELECTRICAL CONTACT

Fig. 8.18 Principle of operation of a reed relay
(a) No current flowing in coil, strips separated
(b) Current flowing in coil, strips are magnetized and attract each other to form an electrical contact

If now two thin strips of material that can be magnetized are placed inside the coil, the strips will become magnetized when the current is flowing in the coil. If the two strips are placed so that one end of each overlaps the other, they will have opposite magnetic polarities and so will attract each other, as shown in Fig. 8.18.

These two strips can be used to form a switch in another electrical circuit. The two strips are placed inside a glass envelope containing an inert gas, and the overlapping portions are coated with gold to give a good reliable electrical contact. The whole assembly contained by the glass envelope is called a *reed insert*, since it is placed inside the electromagnet coil.

A typical reed relay, as illustrated in Fig. 8.18, has *four* of these reed inserts placed inside the electromagnet coil, and each of these can be used to switch a separate electrical circuit.

9 Introduction to Radar and Navigational Aids

Principles

When a radio wave strikes an object or target, some of the energy is reflected or re-radiated by the target back towards the transmitting aerial, so that the presence of the target is *detected*. If the transmitted radio wave is in the form of pulses, then by measuring the time delay between the transmitted pulse and the received pulse or echo, the distance of the detected object from the transmitting aerial can be determined, since the velocity of propagation of radio waves in the atmosphere is constant at the speed of light (3×10^8 metres per second), and distance = velocity × time.

By using *highly directional* transmitting aerials that send a radio wave along a very narrow beam, it is possible also to identify the bearing or direction of the distant target relative to the transmitter.

Thus, we get a system of RAdio Detection And Ranging, which prompted the introduction of the word RADAR. The first important point to make is that this system can only *provide* information, and does not carry information, as has been the case in all telecommunication systems previously considered.

The radar system briefly described above is known as a PULSED PRIMARY RADAR SYSTEM. The radar transmitting aerial, as stated, is highly directional, and there are two main directions to be considered.

The first is the *horizontal* direction relative to the aerial. This is called the AZIMUTH, and the *beamwidth* of the aerial in this plane is very narrow—just a few degrees. The second is the direction upwards relative to the horizontal, that is the *vertical* direction or ELEVATION. This can be quite broad beam to accept reflected or echo signals from a wide range of vertical angles.

(a)   HORIZONTAL SCAN ONLY

Path of aerial scan

(b)   VERTICAL AND HORIZONTAL SCAN TOGETHER

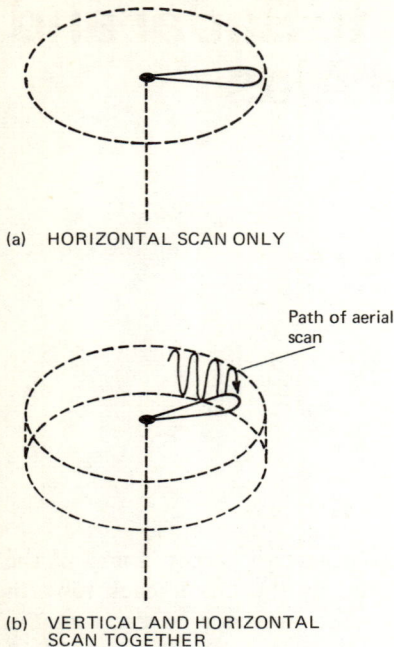

Fig. 9.1 Two methods of narrow-beam scanning

If this aerial is physically turned through 90° so that the original horizontal and vertical planes are interchanged, it will then give a narrow elevation beamwidth, with a broad azimuth beamwidth.

Another possibility is an aerial with narrow azimuth beam-width and multiple narrow elevation beams, which can give a three-dimensional coverage of range, bearing and height, if the aerial is arranged to rotate or scan, on its own axis, in the horizontal plane. (See Fig. 9.1.)

Measurement of Range or Distance

The principle of measuring the RANGE or distance of a target is shown in Fig. 9.2.

The transmitter produces PULSES of radio waves in the frequency range 150 MHz to 30 000 MHz (or 150 MHz to 30 GHz). The *duration* of the pulses will generally be between 0.25 and 50 microseconds (μs). The single aerial is usually used for transmitting and receiving pulses, and the *combining unit* shown in Fig. 9.2 isolates the sensitive receiver from the high-power transmitter pulses, and then switches the aerial to the receiver during the intervals between transmitter pulses. The *time interval* between pulses depends on the maximum distance at which the radar system is to be effective.

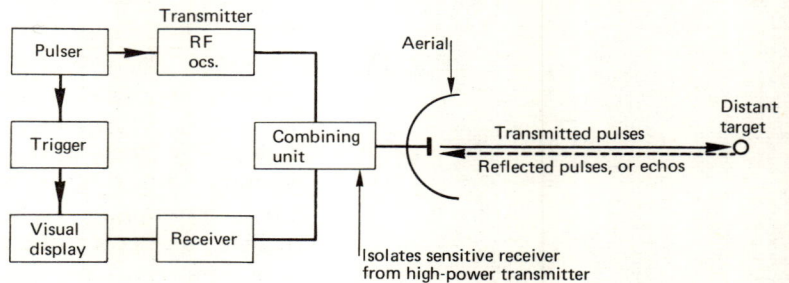

Fig. 9.2 Principle of measuring the range or distance

The transmitted and received pulses are displayed on a cathode-ray tube with a horizontal scan to give a *visual* indication of the time delay. The scan of the CRT display is synchronized with the transmitted pulses, and the transmitted and received pulses deflect the scan vertically to give a visual indication of each pulse.

Because of the constant velocity of propagation of the radio pulses, the time delay can be directly calibrated in distance. The time for the echo pulse to return to the receiver gives *total* distance out to the target and back again. This is illustrated in Fig. 9.3 and is called an A-SCOPE or A-scan display. Because the pulses deflect the scan on the CRT

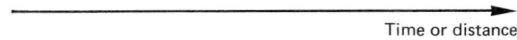

(a)   HORIZONTAL SCAN WITH NO SUPERIMPOSED PULSES

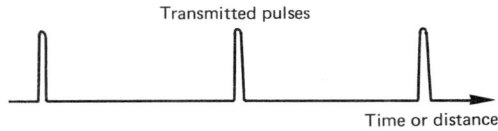

Transmitted pulses

(b)   HORIZONTAL SCAN WITH TRANSMITTED PULSES ONLY

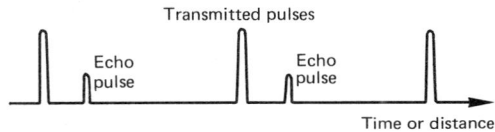

Transmitted pulses

Echo pulse

Echo pulse

(c)   HORIZONTAL SCAN WITH CLOSE TARGET

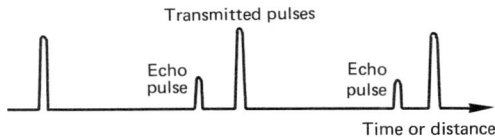

Transmitted pulses

Echo pulse

Echo pulse

(d)   HORIZONTAL SCAN WITH DISTANT TARGET

Fig. 9.3 Principle of vertical deflection A-scan display

vertically from its normal horizontal path, this type of visual indication is called DEFLECTION MODULATION.

The transmitted pulses are actually short pulses of a high-frequency carrier, as shown in Fig. 9.4. For example, if the radio frequency is 1 GHz (1000 MHz), the time for one cycle (periodic time) is

$$T = \frac{1}{f} = \frac{1}{1000 \times 10^6} \text{ sec} = \frac{1}{1000} \, \mu s$$

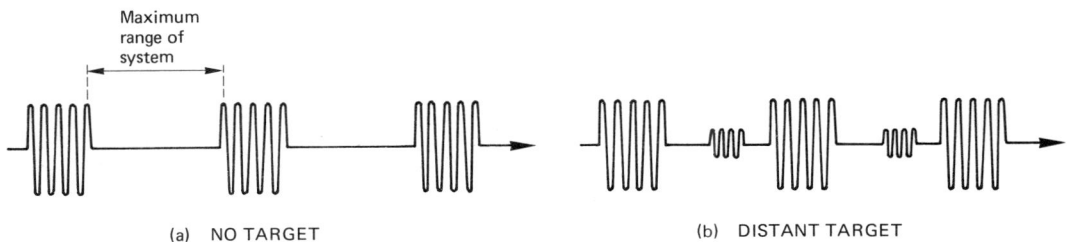

Maximum range of system

(a)   NO TARGET

(b)   DISTANT TARGET

Fig. 9.4 Actual form of pulses

So, if the length of the pulse is 1 μs, there will be 1000 cycles of carrier frequency in each pulse. If the time interval between pulses is 1000 μs, then the maximum effective range will be

$$\text{Velocity} \times \text{time} = (300 \times 10^6 \, \text{m/s} \times 1000 \times 10^{-6} \, \text{s}) \div 2$$
$$= 150 \, \text{km} = 93.75 \, \text{mile}$$

Note: the calculated time must be divided by 2 as shown because the pulse actually travels a distance of 300 km, to the target and back.

Measurement of Direction or Bearing

Having described a method of obtaining the distance or range of a target, we will now consider how to obtain direction or BEARING of the target as well.

The highly directional aerial is rotated so that it searches over the whole of the horizontal plane during each revolution, which usually takes about $\frac{2}{3}$ second. As the aerial rotates, its bearing or direction relative to *north* for *azimuth* (and also relative to *horizontal* for *elevation* in some systems) is converted into either an analogue or a digital electrical signal to apply to the visual display system.

The azimuth or horizontal bearing relative to north is displayed by a moving electron spot that traces out a narrow rotating beam on a circular cathode ray tube which has a long afterglow. The rotation of the trace is synchronized with the aerial rotation. The simple principle of this arrangement, which is called a PLAN POSITION INDICATOR (PPI), is illustrated in Fig. 9.5.

Any reflected or echo signal received from a target *brightens* the spot in order to give a visual indication of the range (by the distance along the scan from the centre) and the bearing (from a calibrated compass scale) in the horizontal plane. A block diagram of this simple pulsed primary radar system is given in Fig. 9.6. Since the receiver echo pulses brighten the trace to give visual indication of distance and bearing, this system is called INTENSITY MODULATION.

For a pulsed primary radar system fitted on board a ship, the "north" position corresponds to the bow of the ship. In the case of a ground pulsed primary radar system tracking aircraft that are close together, it is necessary to request individual aircraft by radio link to carry out a procedural turn to left or right in order to identify each one. This is time-consuming, expensive in fuel, and increases the load on the radio communication equipment at the control point.

Any moving target that is to be identified in range and bearing will be displayed amongst a *clutter* of stationary "blips" produced by reflections from any buildings and other objects scanned by the rotating aerial.

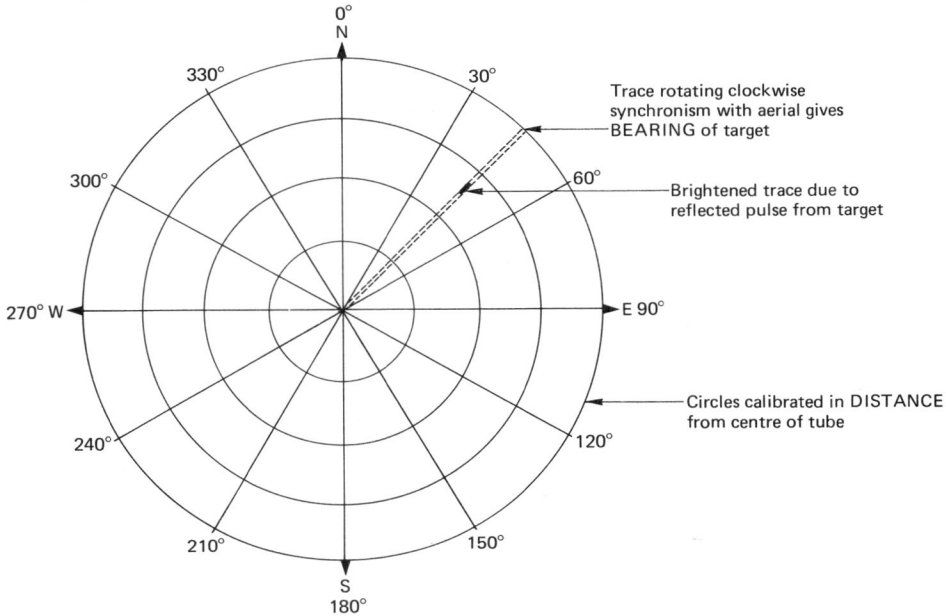

Fig. 9.5 Principle of Plan Position Indicator intensity display of target bearing

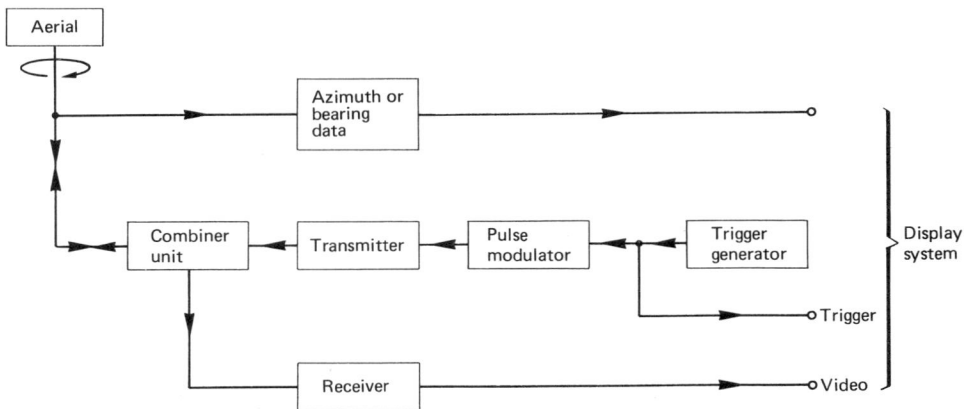

Fig. 9.6 Block diagram of simple pulsed primary radar with PPI display of target bearing

Secondary Radar

This is the name given to a more sophisticated system which gives the *identification* as well as the position of a target.

The target contains a radar transmitter/receiver or TRANSPONDER. The receiver detects the pulse transmitted from the control at a particular frequency and triggers the associated transmitter in the target transponder to transmit a pulse back to control. This gives a *stronger* echo signal at the control than for a purely reflected wave. Typical frequencies would be

Control to target, 1030 MHz (1.03 GHz)
Target to control, 1090 MHz (1.09 GHz)

Further, by using *coded* signals it is possible for control to address a particular target (interrogation) and for that target to give information on identity and height by coded signals (response). Typical interrogation codes are given below:

Interrogation Code	Pulse Spacing (microseconds µs)	Use
1	3	Military identity
2	5	Military identity
3/A	8	Joint military/ civil identity
B	17	Civil identity
C	21	Altitude
D	25	Unassigned

Since the control display will only respond to predetermined re-transmitted signals from a definite target, the problem of clutter is greatly reduced, since unwanted stationary objects will *not* re-radiate a strong signal at an appropriate frequency.

The main advantages of secondary radar over primary radar systems can be summarized as

(1) Larger echo signal at the control receiver.
(2) Identification of target as well as its position.
(3) Targets addressed only when needed by control.
(4) Variety of information possible from targets.
(5) 'Clutter' greatly reduced, giving a moving-target indicator system (MTI).

Radar can be employed in several ways as a navigational aid for civil use and for several military uses.

CIVIL USES

(1) Since mountains, plains, cities, rivers, oceans, etc. all reflect radio waves to different extents, radar equipment in aircraft can provide information about the ground below despite poor visibility or darkness.

(2) Radar equipment on a ship can similarly give information on the location of other ships, marker buoys, land, etc. despite poor visibility or darkness.

Also, radar systems can be used to

(3) Give aircraft and ships information as to their exact position at any moment.

(4) Assist aircraft in landing during poor visibility or darkness.

(5) Enable aircraft to measure height above ground.

(6) Enable air space above an airport to be monitored to control take–off and landing of aircraft.

(7) Check speed of vehicles on motorway.

MILITARY USES

(1) Aiming ground-based guns at ships and aircraft.

(2) Locating ground targets for bomber aircraft through clouds or at night.

(3) Locating moving targets for fighter aircraft at night.

(4) Directing guided missiles from gound, ships or aircraft.

(5) Tracking enemy missiles or aircraft to give early warning systems.

(6) Searching for submarines.

Radio Navigational Systems

Radio waves have long been used as an aid to navigation.
Some of the many different systems will now be discussed.

Visual–Aural Range (VAR)

This is a ground-based system which enables an aircraft to determine its position relative to the ground transmitter.

Four separate signals are transmitted at the same radio frequency from directional aerials. Two of the transmissions are in the form of Morse Code signals, one representing letter N (dash dot) and the other representing letter A (dot dash). The other two carriers are modulated by 90 Hz and 150 Hz tones respectively.

In an aircraft, the receiver demodulates the four signals, converting the N and A signals into *audible* tones for the pilot, and converting the 90 and 150 Hz tones into d.c. signals which indicate blue or yellow respectively on an instrument. The waveform and directivity of the four transmitted signals are illustrated in Fig. 9.7.

There is a range area around the four transmitting aerials in which the signals are workable, and the range area is divided into four quadrants by the directional aerials. This is illustrated in Fig. 9.8. Each quadrant contains a particular *pair* of signals, which enables the pilot in an aircraft to determine which quadrant he is in by the two demodulated signals received. For example

Quadrant 1 – yellow and N
Quadrant 2 – yellow and A
Quadrant 3 – blue and A
Quadrant 4 – blue and N

Along the line separating the Blue and Yellow quadrants, the d.c. outputs produced by the demodulation of the 90 Hz and 150 Hz tones cancel so that the receiver instruments indicates neither blue nor yellow.

Along the line separating the N and A quadrants, the receiver output gives a continuous audible signal, instead of either N or A notes.

Visual Omnirange (VOR)

This system enables the pilot of an aircraft to read an exact bearing on a single instrument. Two signals are radiated from the ground transmitter. One signal, called the *reference* signal, is obtained by a 30 Hz tone frequency-modulating a 9960 Hz sub-carrier, and the resulting f.m. wave then amplitude-modulates a radio frequency carrier. This signal is radiated in all directions from the transmitter.

The second radiated signal is obtained by a 30 Hz tone amplitude-modulating the carrier, whose phase is varied according to the *direction* of radiation from the transmitter. The two radiated signals are *in phase* for a direction due *south* of the transmitter, and the phase difference between the two signals increases for a clockwise direction from due south. The radio frequency carrier is usually in the range 112 MHz to 118 MHz.

Using a sub-carrier for the reference signal enables the two signals to be separated by a filter at the receiver. The two demodulated 30 Hz tones can be compared in phase, and the phase difference is used to indicate an exact bearing of the aircraft from the transmitter.

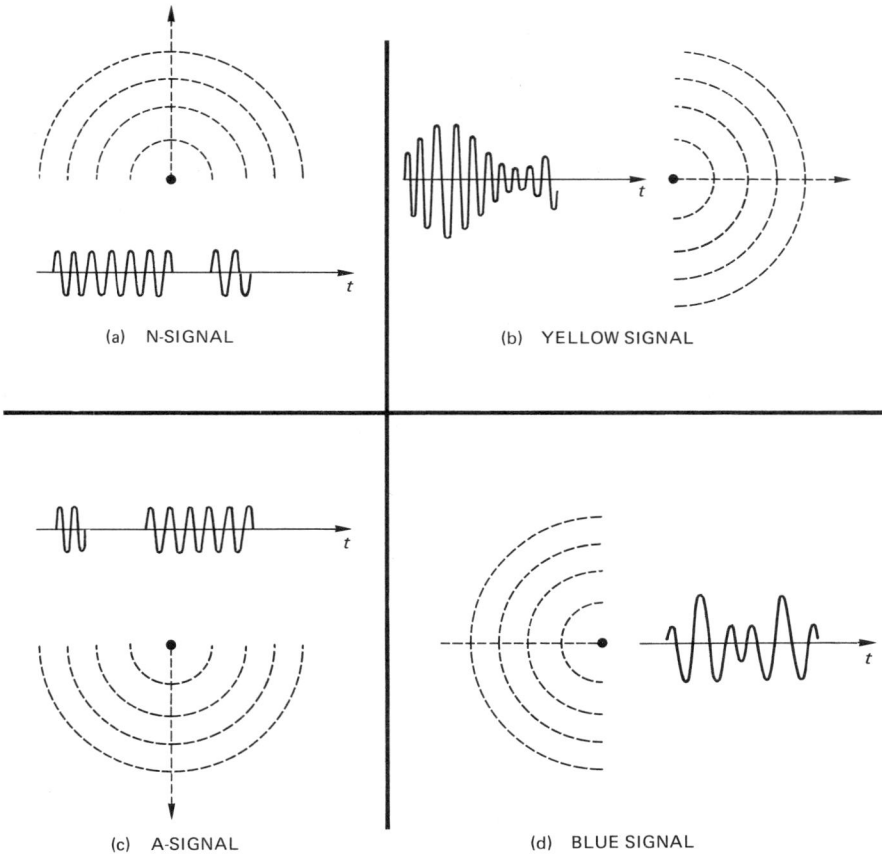

(a) N-SIGNAL

(b) YELLOW SIGNAL

(c) A-SIGNAL

(d) BLUE SIGNAL

Fig. 9.7 The four directional signals radiated in VAR

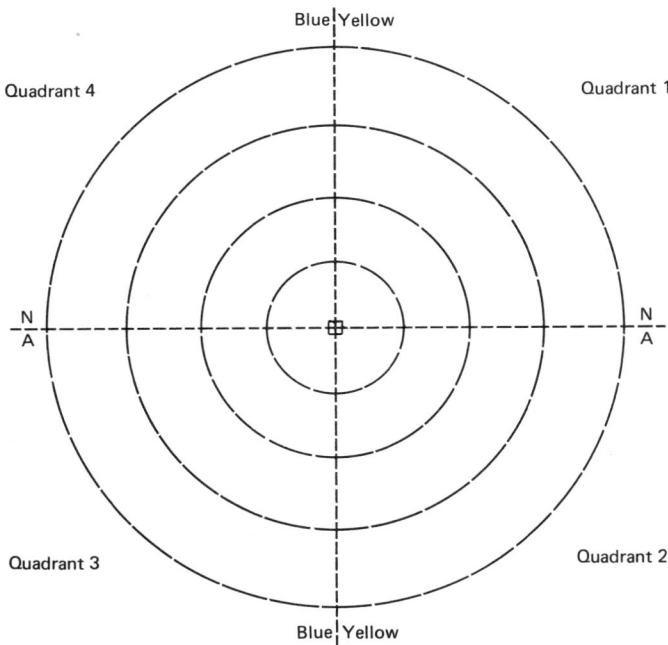

Fig. 9.8 Transmitted wave from the four VAR aerials

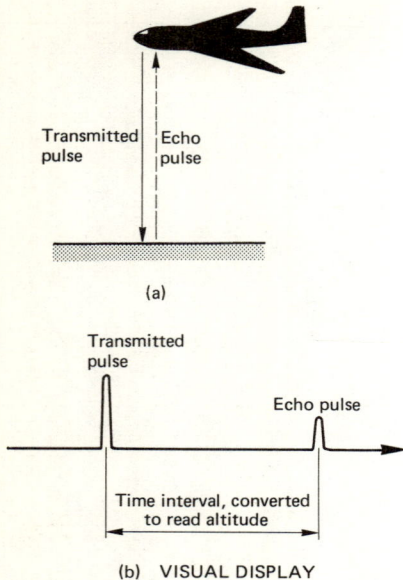

(a)

(b) VISUAL DISPLAY

Fig. 9.9 Simple principle of radio altimeter

Radio Altimeters

An aircraft transmitter sends a signal down to the ground, and the time taken for a reflected echo pulse to return to the aircraft receiver is given by a visual display and converted into an altitude reading. This is illustrated very simply in Fig. 9.9.

The Decca Navigational System for Ships

This system is based on a *phase-difference* measurement.

The phase of the radio-frequency carrier of a ground wave is delayed by 360° for every wavelength of distance from the transmitting aerial. So, if two fixed transmitting stations radiate carriers having the same frequency that are locked in phase, then at one point that is the same distance from both transmitters, the two received signals will have the same phase.

Similarly, for all points at which there is a constant phase difference between the two signals, there must also be a constant difference of distance from the two stations. If all points having a constant "difference" distance from the two stations are joined together, the line joining the points (or locus) is in the shape of a hyperbola with the two stations as the foci. This is illustrated in Fig. 9.10. Other lines (or loci) can be added to indicate multiples of 360° phase delay,

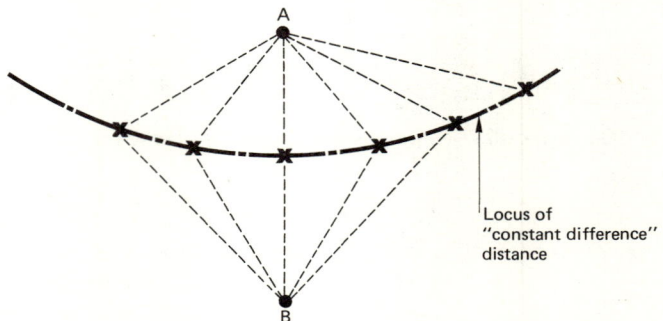

Fig. 9.10 Hyperbolic locus of "constant-difference" distance

This principle is extended in a typical practical system by having a MASTER station and three SLAVE stations. Each slave station forms a pair with the master so that three sets of lines as shown in Fig. 9.10 intersect to give *position fixing* information through the whole 360° of azimuth (horizontal direction from transmitter).

The space between the hyperbolic lines which correspond to adjacent multiples of 360° phase shift is called a *lane*. Several installations as illustrated in Fig. 9.11 are provided at different points along various coastlines in an area (e.g. Western Europe) to give a Decca Navigator Chain.

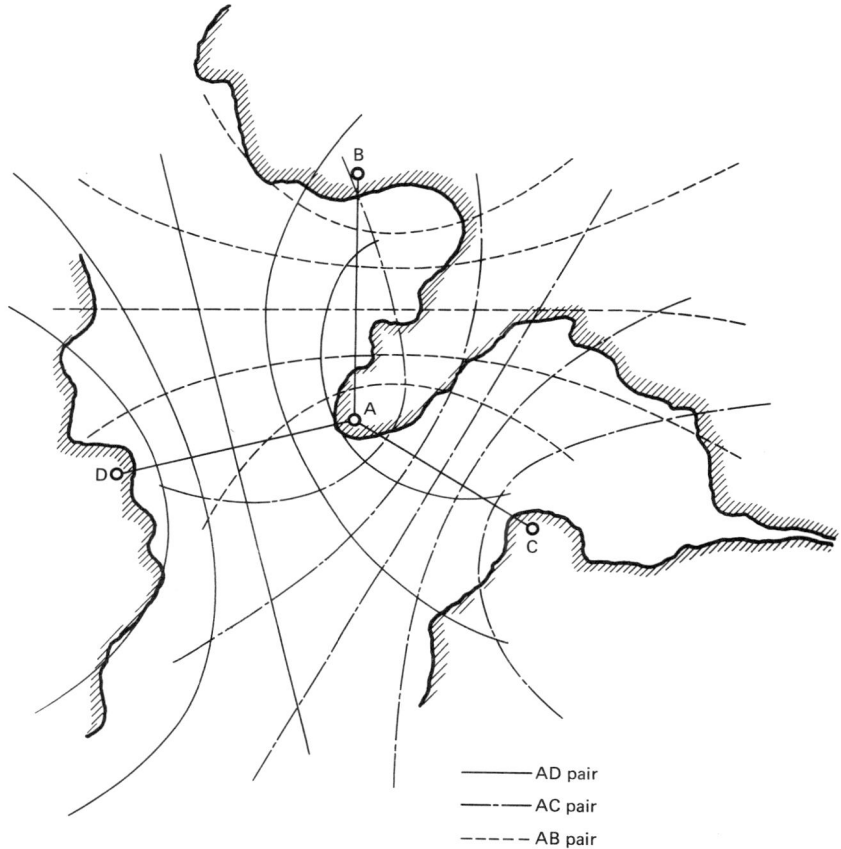

Fig. 9.11 Decca Navigator system with master (A) and three slaves (B, C, D)

——————— AD pair
—·—·— AC pair
— — — — AB pair

Loran System

The LOng RAnge Navigational aid system was developed in the USA during the last war as a long-distance navigational aid for aircraft and ships. This system is based on a *pulse arrival-time difference* measurement as opposed to the phase difference measurement of the Decca system.

Fixed transmitting stations are arranged in pairs. One station of a pair transmits a pulse at a given frequency, and this pulse is received by the other station of the pair. The second station then transmits a pulse on the same frequency as the first station after a definite time delay.

There will be a number of points at which the two transmitted pulses will arrive with a constant time difference, representing also of course a constant path-length difference. As for the Decca system, these "equal difference" points will be on a line or locus that has the shape of a hyperbola, with the two transmitting stations as foci. The pulses are transmitted at a rate of approximately 30 times per second. "Position lines" can be set up by measuring differences in pulse-arrival times.

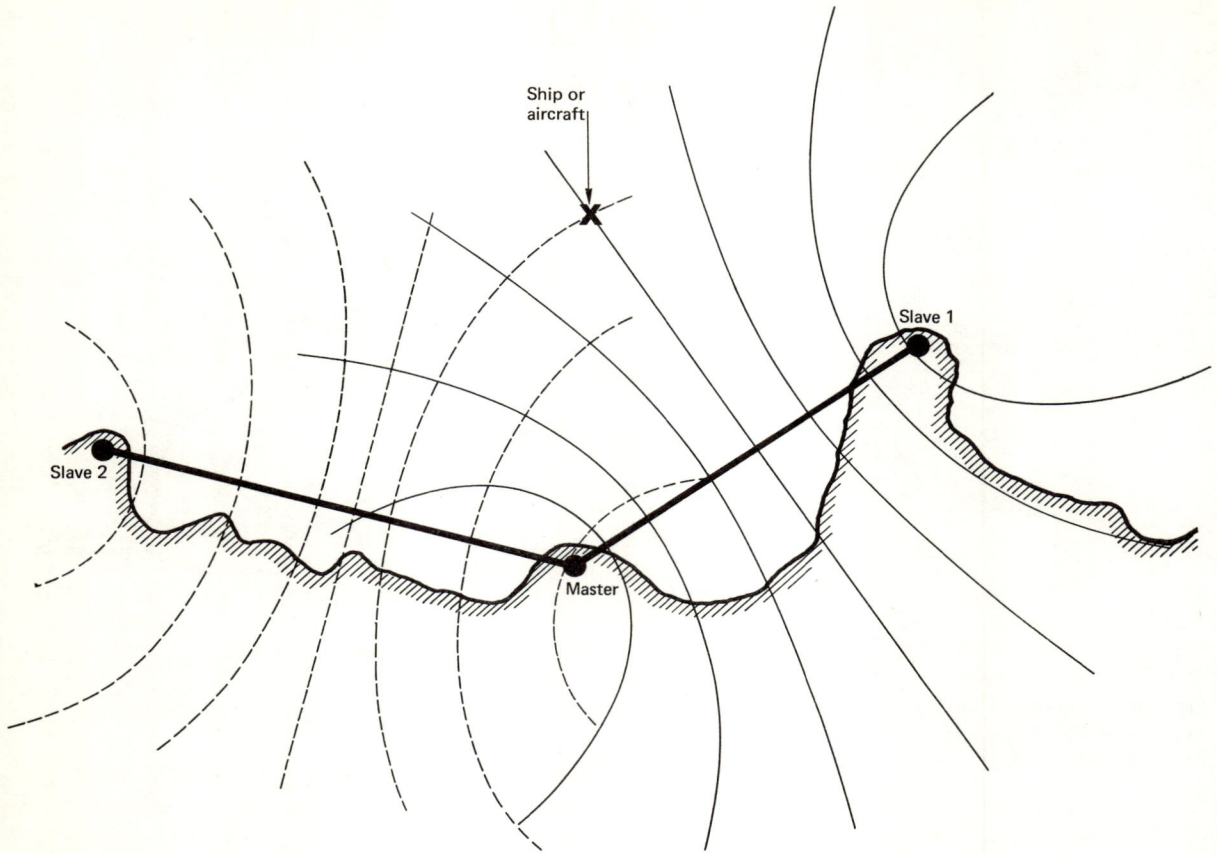

Fig. 9.12 Loran system

The measurement is achieved by a visual display on a CRT, with time bases locked and provided with timing marker "blips" derived from a 100 kHz crystal-controlled oscillator in the receiver.

A second pair of stations is then established to provide another set of "position lines" by transmitting pulses on the same frequency as the first pair. A slightly different pulse recurrence rate is used, however, so that the visual display of the second pair of pulses slowly drifts across the time bases locked to the first pair. Thus the two displays can be distinguished.

Generally, a common control station, or Master, will be used for two pairs, using two slave stations, with a station separation of 320 to 480 kilometres (200 to 300 miles). The radiated frequency is around 2 MHz. Fig. 9.12 illustrates the principle of the Loran system.

10 Introduction to Data Transmission

Use of Computer Systems

In recent years there has been a rapid development in small and medium sized electronic computer systems, as well as large-scale electronic computers, to help and improve the organization and management of industry, commerce, and trade. This was perhaps a logical development from the electro-mechanical machines which have been used for a long time for book-keeping, invoicing, simple mathematical calculations, and so on.

Small and medium sized computer systems have helped to improve communication between people and computers, and have brought computers into the every-day field to help in solving problems and tasks. The very heart of computers has been brought into offices and other places of work instead of being hidden away in some anonymous place, as is usually the case with large-scale computers.

In the early days, problems were brought to a computer by the customer or user, and the results were eventually carried away on a print-out. Obvious developments were to get computers to interact, or work together, and to arrange for access to a computer from a distance. The idea of a *computer network*, on the lines of a telephone or telegraph network, was clearly a desirable objective, with the network eventually handling several different types of information.

One particular type of small/medium sized computer system that has been of particular importance in this respect is visible record equipment. This is called a Visible Record Computer, and clearly has great potential in many different applications. It has also proved possible and advantageous to use these small/medium sized systems as input/output stations for centralized large-scale computers, which have high programming ability, large storage facilities, and automated data processing ability.

CRT — Cathay Ray Tube

This technological development has made it possible to transmit information rapidly between two points, with a visual display if needed, and a facility for storing information until required. This technique of transmitting information by use of electronic devices is called DATA COMMUNICATION or Data Transmission.

Before going any further it may be useful to introduce some of the terms used in connection with computers and data communication.

HARDWARE   This term refers to the physical equipment that makes up a computer system, that is the mechanical, magnetic, electronic and electrical devices.

SOFTWARE   This comprises the programmes, procedures and associated documentation concerned with the operation of a computer, and includes such items as assemblers, compilers, library routines, etc.

REAL TIME   Real time is the actual time during which a physical event occurs, or more specifically, the time of occurence of physical events in which a computer can be used to control or guide the events.

TIME SHARING   This is a method of using computers that enables a number of users to carry out different programmes apparently at the same time, and allows the computer to interact with these programmes during completion.

PROGRAMME (or PROGRAM)   This is a plan for the solution of a problem by a computer. It consists of a series of instructions or statements in a form acceptable to a computer, and is prepared to achieve a specific result.

INTERFACE   This is a connection or common boundary between any two units or devices.

OFF-LINE   This describes any equipment that is *not* under the direct control of the central processor of a computer. It can also apply to any terminal equipment that is *not* connected to a transmission line.

ON-LINE   This describes any equipment that is under the direct control of the central processor of a computer. It can also apply to any terminal equipment that is connected to a transmission line.

MODEM   This is a contraction of *mo*dulator/*dem*odulator, which is a device used to modulate and demodulate signals at either end of a data communication system, and enables digital signals to be transmitted over an analogue (e.g. speech) network.

TERMINAL   This is any point in a system at which data (information) can enter or leave the system, or any device capable of sending or receiving information over a data system.

DATEL   This is a contraction of *da*ta *tele*communication, and is the term used by the British Post Office to describe the various data transmission facilities available, from low-speed telegraphy to high-speed data systems.

BIT   This is a contraction of *bi*nary digi*t*, and refers to one of the two digits (0 and 1) used in binary notation. It is a single character in a binary number.

BYTE   This is a sequence of binary digits (bits) used together as the smallest addressable unit in a particular memory.

WORD   This refers to a sequence of bits treated as an entity to form a character (longer than a byte) and a basic unit of data in a memory.

Types of Computer

Analogue Computers

This type of computer continuously measures variable physical quantities (e.g. pressure, temperature, fluid flow, etc.) and represents them by some physical analogy, such as voltage or shaft rotation, usually in real time.

Analogue computers are generally classified as

(*a*) *General purpose* (indirect), which provide valid mathematical solutions useful for any problem that can be expressed in mathematical terms.

(*b*) *Fixed purpose* (direct), which solve the problem presented by a specific situation by setting up a direct analogy to the characteristics and parameters of the problem.

Some typical uses for analogue computers are
Prediction of performance of a prototype aircraft design.
Solution of complex mathematical problems.
Plotting the flight of missiles.
Control of industrial and chemical processes, e.g. blending of petrol.

Digital Computers

This type of computer operates on information represented by combinations of discrete signals, and performs arithmetic and logic processes on this information, not necessarily in real time.

Digital computers can be classified as general purpose or special purpose, although the special purpose computers are really offshoots of general purpose types. Some typical uses for digital computers are

Invoicing. 金计, 开发票
Payroll information and preparation.
Book-keeping.
Materials accounting, bank accounting, etc.
Automatic control of machine tools.
Airline ticket reservations.

Hybrid Computers

This type of computer makes use of both analogue and digital techniques, and includes analogue-to-digital and digital-to-analogue converters.

Advantages of Computers

(1) Once programmed, long sequences of operations can be performed without human help, and the necessary answers produced.
(2) Complex mathematical problems can be solved very quickly.
(3) Large quantities of information can be stored.
(4) Very high degrees of accuracy can be achieved.

On the other hand, there are a number of *disadvantages* in using computers, including the following:

(1) They are very expensive to install and maintain.
(2) They require a clean dust-free environment with a relatively constant background temperature.
(3) They require specialist operating and programming staff.
(4) They normally do not question the accuracy or truth of input information, although they can be programmed to accept information only within certain pre-determined limits.
(5) They are not capable of creative or original thought.
(6) They present security problems.

Elements of a Data Communication System

It was stated earlier that the need had arisen for ways of transmitting, receiving and interchanging data (information) rapidly between various points. It is now appropriate to consider some of the ways in which information can be presented in one place in order to be reproduced at another place.

Most systems start with a source document which has to be changed into a form acceptable to a computer, either by an operator working a keyboard or by a suitable machine.

Input Devices

(1) TELEPRINTER
The simple principle of the teleprinter has been considered at the beginning of Chapter 7. Briefly, by selecting a written message by means of a keyboard, each character, letter or figure is represented by a series of positive and negative voltage pulses (± 80 V) in accordance with a predetermined code (Murray code).

(2) PUNCHED CARD
The information is manually or automatically punched on to a card by a pattern of holes in accordance with a code. Typically, a punched card is $7\frac{3}{8}$ in. $\times 3\frac{1}{4}$ in. (18.8 cm $\times$ 8.3 cm), containing 80 vertical columns in which the information or data can be placed.

(3) PUNCHED TAPE
This is usually paper tape on which a pattern of holes is punched in lines across the tape to represent data. Each line across the tape can use either 5, 6, 7 or 8 holes, according to the system being used.

(4) MAGNETIC TAPE
A thin metal or plastic tape is coated with a magnetic surface on which data can be stored as small magnetized spots.

Output Devices

Each of these four devices which produce a particular form of *input* to a data communication system from a source document can also be used as *output* devices at the other end of the system.

In addition, the output can be presented as a visual display on a cathode ray tube, either as letters and/or figures or in a graphical form. This is called a *Visual Display Unit* (VDU), and often has a keyboard associated with it for producing an input signal.

It was stated earlier that any device capable of sending or receiving information over a data system is called a TERMINAL. In order to connect the terminal to another terminal or central computer, a telephone-type line is generally used. In order that the line can handle the data signal, a MODEM has to be connected between the terminal and the line, as illustrated in Fig. 10.1. The terminals can be permanently connected by an *exclusive rented* line, or the connection can be made over the public switched telephone network (p.s.t.n.).

Fig. 10.1 Simple data communication system

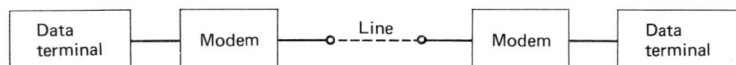

The Binary Number System

We should now look at the type of signal that is used in data communication or transmission. It is based on the BINARY system of counting, which has only *two* values (0 and 1) instead of ten as in the decimal or denary system of counting (0, 1, 2, 3, 4, 5, 6, 7, 8, 9). (Note: bi means 2.) Table 10.1 gives a comparison of these two number systems.

Table 10.1 Comparison of denary and binary counting

Denary	0	1	2	3	4	5	6	7	8	9	10
Binary	0	1	10	11	100	101	110	111	1000	1001	1010
Denary	11	12	13	14	15	16	17	18	19	20	etc.
Binary	1011	1100	1101	1110	1111	10000	10001	10010	10011	10100	etc.

Now, in the denary system, using a base of 10, we see that 4857 really means

$$4000+800+50+7$$

or

$$4\times10^3+8\times10^2+50\times10^1+7\times10^0$$

(Remember that any quantity to the power zero is 1.)

In the same way, in the binary counting system, using a base of 2, we can see that 10100 represents

$$1\times2^4+0\times2^3+1\times2^2+0\times2^1+0\times2^0$$

or

$$1\times16+0\times8+1\times4+0\times2+0\times1$$
$$=16+\ 0\ +\ 4\ +\ 0\ +\ 0\ =20$$

Table 10.1 shows that

(*a*) 1 binary column is needed for denary numbers 0 and 1.
(*b*) 2 binary columns are needed for denary numbers 2 and 3.
(*c*) 3 binary columns are needed for denary numbers 4, 5, 6, 7.
(*d*) 4 binary columns are needed for denary number 8 to 15.

However, we can use 4 binary columns to represent denary numbers from 0 to 15 (i.e. sixteen numbers altogether) by placing zeros in the unused columns, e.g.

Denary 1 is represented by 0001.
Denary 5 is represented by 0101. (etc.)

Generally, for n binary columns, we can represent 2^n denary values or numbers:

for $n = 1$,

$2^1 = 2$ denary numbers can be represented $(0, 1)$

for $n = 2$,

$2^2 = 4$ denary numbers can be represented $(0, 1, 2, 3)$

for $n = 3$,

$2^3 = 8$ denary numbers can be represented $(0, 1, 2, 3, 4, 5, 6, 7)$

for $n = 4$,

$2^4 = 16$ denary numbers can be represented $(0 \text{ to } 15)$

for $n = 5$,

$2^5 = 32$ denary numbers can be represented $(0 \text{ to } 31)$

and so on.

Now, most electrical or electronic components have only two stable states, that is

ON or OFF

CURRENT or NO CURRENT

VOLTAGE or NO VOLTAGE

These two stable states, or BISTABLE conditions, can be represented in the binary system by associating one state with 0 and the other state with 1. Some simple examples are shown in Fig. 10.2.

Each single binary condition, either 0 or 1, is called a binary digit or BIT. A series of bits can therefore represent a particular denary number as shown in Table 10.1. The number of bits used or transmitted in one second represents the *binary signalling speed* in bits per second or bit/s.

We have already seen that a teleprinter produces a basic signal that consists of either $+80\,V$ or $-80\,V$ pulses. This is a form of binary signal with $+80\,V = 0$ and $-80\,V = 1$. The pulses are arranged in a 5-unit code, with each bit lasting 20 ms, giving a maximum rate of 50 bit/s or 50 baud. (Baud was the name used in early telegraphy.)

There are several different types of data communication systems and modems using different signalling speeds, such as 110, 200, 600, 2400, 4800 and 48 000 bit/s, over the public switched telephone network, or over private rented lines. In the future, systems will use signalling speeds up to 4.8×10^6 bit/s (4.8 Mbit/s).

(a) RELAY WITH ONE CONTACT

       (i)   Not energized = 0   (ii) Energized = 1

(b) TRANSISTOR

       (i)   Non-conducting = 0   (ii) Conducting = 1

(c) ELECTRO-MAGNET

       (i)   Negative saturation = 0   (ii) Positive saturation = 1

(d) VOLTAGE LEVEL

High level = 1
Low level = 0

(e) CURRENT PULSES

       (i)   No pulse = 0   (ii) Pulse = 1

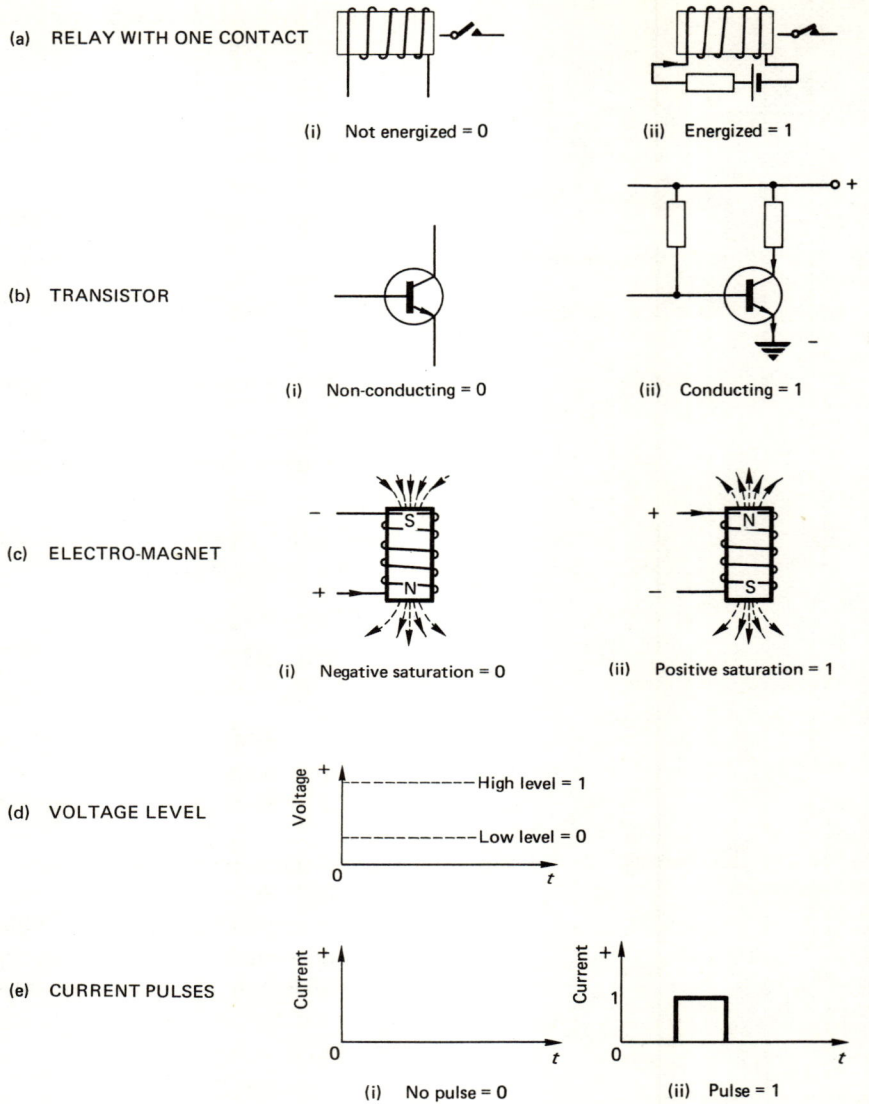

Fig. 10.2 Some simple electrical and electronic binary devices or states

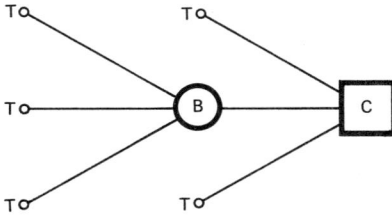

Fig. 10.3 Simple principle of multi-point data network (C-central computer, T-data terminal, B-branching point)

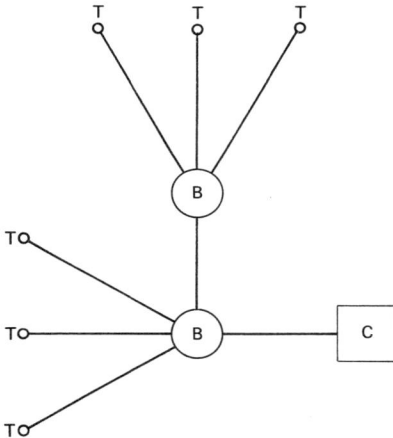

Fig. 10.4 "Double-star" multi-point data network

Data Network

Fig. 10.1 illustrated the simplest form of data network containing two terminals connected together over a dialled p.s.t.n. link or an exclusive private line. A number of business organizations have established multi-point networks linked to a central computer centre to give access to a variety of data information with efficient use of lines. The principle of these multi-point networks is illustrated in Fig. 10.3. This simple network can be extended to include further "branching points" or connections as shown in Fig. 10.4.

In Figs. 10.3 and 10.4 the branching points simply allow a number of terminals to share a single line to the central computer, without any change in signalling speed.

Another method of concentration is to use a MULTIPLEXER arrangement to take a number of slow-speed inputs from separate terminals and apply them to a shared single link at a faster speed. This is illustrated simply in Fig. 10.5.

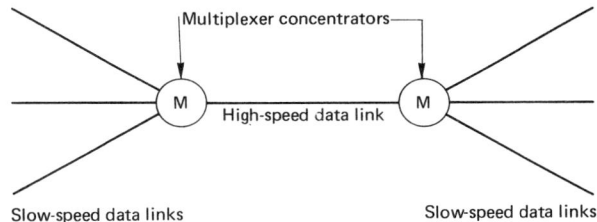

Fig. 10.5 Use of multiplexer concentrators

Polling

When a number of terminals share the same central computer, the computer can only deal with one terminal at a time, so it is arranged for the computer to interrogate each terminal in turn to see if it needs a service. This is called POLLING.

Alternatively, if multiplexers are used in a network, it can be arranged for each multiplexer to poll its own particular terminals. This gives less polling direct to the central computer and so increases overall efficiency of the network.

The simple arrangement shown in Figs. 10.3, 10.4 and 10.5 require that the central computer carries out polling, sorts out messages, and so on. Many tasks in a large data network are simple and repetitive, yet can take up a lot of computer time. It is sometimes more economic if these simple repetitive functions are carried out by small units called Front End Processors (FEP), as illustrated in Fig. 10.6.

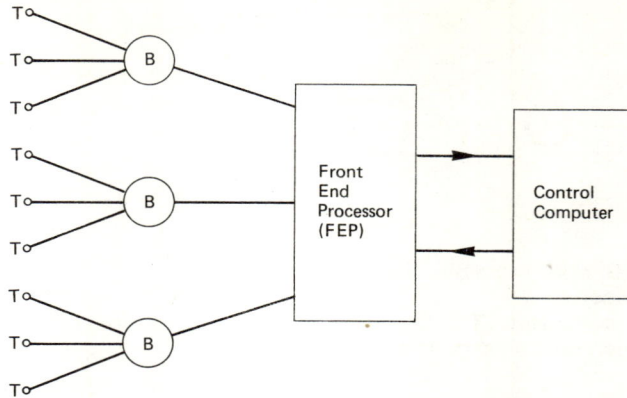

Fig. 10.6 Use of front end processor (FEP)

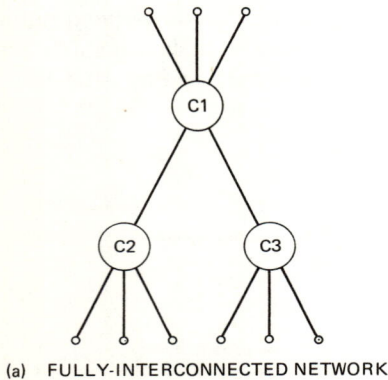

(a) FULLY-INTERCONNECTED NETWORK

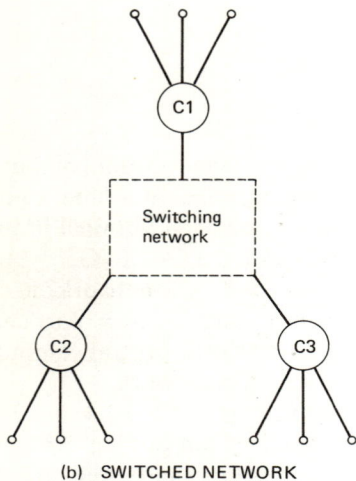

(b) SWITCHED NETWORK

Fig. 10.7 Large data network using more than one control computer

Very large organizations may find it an advantage to use more than one computer to process the network data, in order perhaps to divide the data into particular functions, e.g. clerical tasks, stock control, pay roll, etc. It will then be necessary to allow the computers to interact with each other to obtain the specific data needed by a particular terminal. Computers can be inter-connected in a number of ways, two of which are shown in Fig. 10.7.

One very important application of a nation-wide computer-controlled data network is that used by the large Banks, which each have their own business to conduct and also are linked by a common Bankers Clearing Service. Generally, the different branches of a Bank can be graded as Major or Minor, with several minor branches each connected to a major branch by a low or medium speed link in order for data to be concentrated before being passed to the central computer over a medium or high-speed link. This is illustrated in Fig. 10.8.

Simple Digital Computer Principles

We now look briefly at the computer itself to see the essential units it needs to carry out its tasks.

It should already be fairly obvious that *inputs* and *outputs* are required, and need to be selected one at a time by a *control unit*. It will usually be necessary to store information in a *memory* device and also on occasions to have access to a large *back-up store* for other information. The control device also obtains the selected input information from the memory, and then acts on the information by instructing an *arithmetic unit* to carry out the necessary operations, before passing the data to the selected output.

The inputs can be of different types, e.g., punched card, punched tape, magnetic tape, etc., so an *instruction decoder* may be needed for the control to understand the instructions.

The arrangement is illustrated simply in Fig. 10.9.

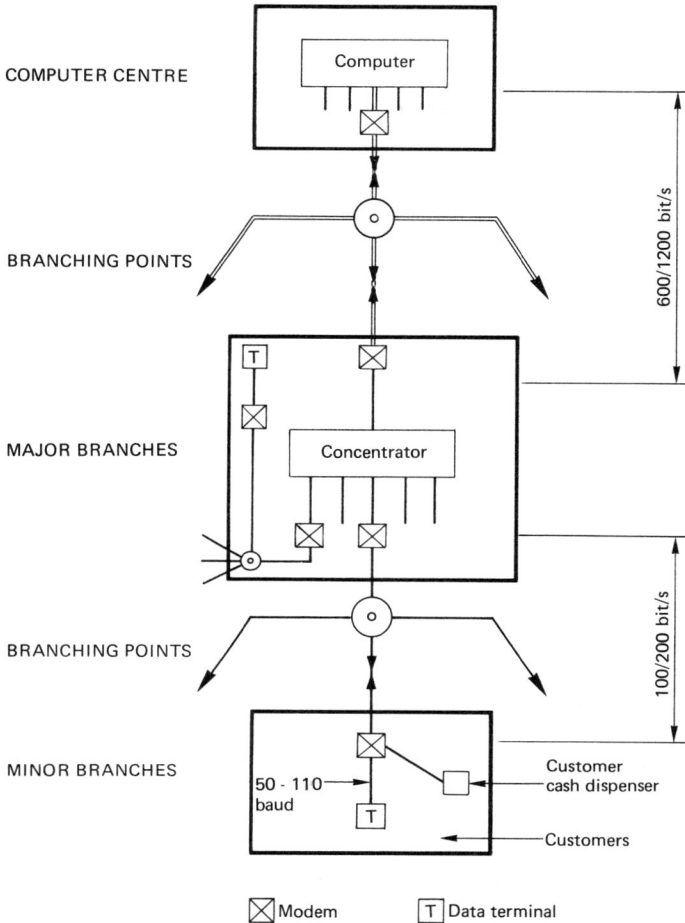

Fig. 10.8 Typical national Bank accounting data network

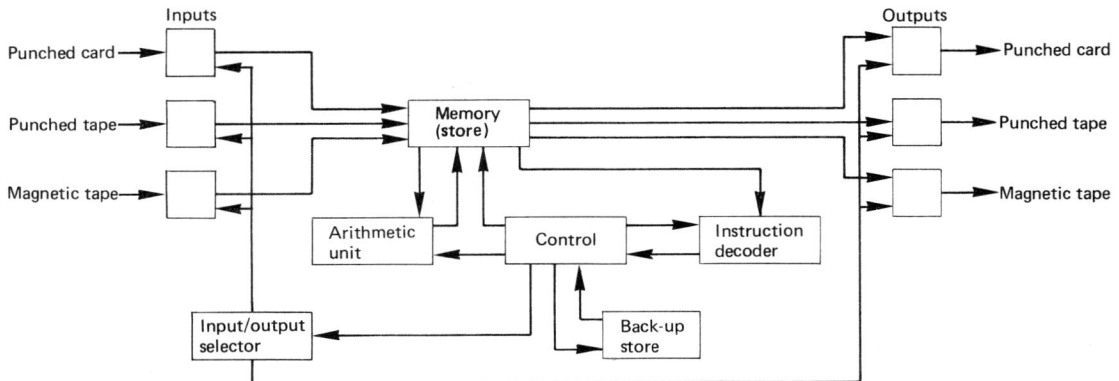

Fig. 10.9 Block diagram of simple digital computer

Telecommunication Systems I: Learning Objectives (TEC)

(A) Information Transmission

(1) Describes how electronics is used to convey information.

(2) Describes modulation.

page 16 2.3 States that modulation is a process whereby some characteristics of a carrier wave are varied by another wave.

17, 19 2.4 Identifies from given waveform diagrams:
    (*a*) an unmodulated carrier
    (*b*) an amplitude-modulated carrier
    (*c*) a frequency-modulated carrier
    (*d*) a pulse-modulated carrier.

17, 19 2.5 Draws and labels pulse and sine-wave amplitude-modulated waveforms.

16 2.6 States that de-modulation is the reverse process of modulation.

14 2.7 States the frequency range associated with speech modulation.

20 2.8 States the bandwidth of an amplitude-modulated waveform as the frequency range between the lowest lower sideband frequency and the highest upper sideband frequency.

16 2.9 States that frequency modulation is the technique of varying the frequency of a carrier wave.

(B) Radio

(3) Understands the basic properties of radio waves and their use in telecommunications.

22 3.1 Draws a block diagram of a commercial broadcast system showing relevant waveforms.

21 3.2 Explains that the path between transmitter and receiver can consist of
    (*a*) cable
    (*b*) space

21 3.3 States that radio waves can be propagated through a vacuum and through insulating materials and are reflected by conducting surfaces.

     3.4 Draws the path taken by
22   (*a*) low radio frequencies, i.e. ground wave
23   (*b*) high radio frequencies, i.e. wave in ionosphere
24   (*c*) very high radio frequencies, i.e. space wave.

25 3.5 Explains that to establish a two-way radio telephony circuit, two transmitters, two receivers and two different frequencies are required.

26 3.6 Draws a block diagram of a two-way radio telephony system.

(C) Television

(4) Describes the cathode-ray tube and the make-up of a monochrome television picture.

(D) Radar and Navigational Aids

(5) Describes how primary radar can be used for range and bearing measurements.

(6) Describes how radio waves can be used for navigation.

(E) Telephony and Telegraphy

(7) Describes a public telephone and telegraph system in functional block form.

(8) Understands the concept of a matrix switch.

(9) Explains how a call is routed through a simple step-by-step exchange.

page 72, 73 9.3 Describes the use of group and final selection to select one from ten.

75 9.4 Labels a simple trunking diagram of an exchange to show how a call is connected.

(10) Understands how speech is transmitted between two telephone subscribers.

38–41 10.1 Explains how speech is converted into electrical energy by means of a simple carbon granule transmitter (the detailed construction of the transmitter is not needed).

43–45 10.2 Explains how electrical energy is converted into sound waves by means of a rocking armature receiver (the detailed construction of the rocking armature receiver is not needed).

45 10.3 Draws a simple diagram of two LB telephones interconnected by a pair of wires.

(11) Explains the need for amplification of signals alone line links.

50–52 11.1 Describes the concept of electrical noise and its relation to signal strength.

49 11.2 Explains that a signal is attenuated due to losses in cable.

53 11.3 Explains the need to amplify the signal at regular intervals along the line.

(F) Data Transmission

(12) Understands the forms of data transmission.

90, 91 12.1 Explains simply the following terms used in data communications: hardware, software, real-time, time sharing, program, interface, off-line, on-line, modem, terminal, datel, bit, byte, word.

91 12.2 Names the two types of computers:
(*a*) analogue
(*b*) digital

91, 92 12.3 States at least two uses of analogue and digital computers.

92 12.4 States the advantages and disadvantages of using computers.

98, 99 12.5 Describes, by means of a block diagram, a data communication system, e.g. a payroll or bank accounting system.

95 12.6 Explain why a two-state code is used in data transmission systems.

page 93 12.7 Describes the following input and output devices:
        (*a*) VDU
        (*b*) teletype
        (*c*) punched card
        (*d*) punched tape
        (*e*) magnetic tape
93 12.8 States typical uses of the devices in 12.7.

Index